Cane River Bohemia

Cammie Henry works on her scrapbooks in her bedroom, 1930.

(Melrose Collection, SB 203, 32, Cammie G. Henry Research Center, Watson Memorial Library, Northwestern State University of Louisiana)

Cane River Bohemia

CAMMIE HENRY AND HER CIRCLE
AT MELROSE PLANTATION

Patricia Austin Becker

LOUISIANA STATE UNIVERSITY PRESS

BATON ROUGE

Published by Louisiana State University Press
Copyright © 2018 by Louisiana State University Press
All rights reserved
Manufactured in the United States of America
First printing

DESIGNER: Michelle A. Neustrom
TYPEFACES: Whitman, text; Meltow San 200
PRINTER AND BINDER: Sheridan Books, Inc.

LIBRARY OF CONGRESS CATALOGING-IN-PUBLICATION DATA

Names: Becker, Patricia Austin, 1959– author.
Title: Cane River Bohemia : Cammie Henry and her circle at Melrose Plantation /
    Patricia Austin Becker.
Description: Baton Rouge : Louisiana State University Press, [2018] | Includes bib-
    liographical references and index.
Identifiers: LCCN 2018012796| ISBN 978-0-8071-6982-7 (cloth : alk. paper) |
    ISBN 978-0-8071-7027-4 (pdf) | ISBN 978-0-8071-7028-1 (epub)
Subjects: LCSH: Henry, Carmelite Garrett, 1871–1948. | Melrose Plantation (La.)
    —History. | Natchitoches Parish (La.)—Biography. | Plantations—Louisiana—
    Melrose Plantation—History.
Classification: LCC F377.N4 B43 2018 | DDC 976.3/6506092 [B] —dc23
LC record available at https://lccn.loc.gov/2018012796

For my parents
Mary Earle Texada Phillips and William "Hutch" Phillips Jr.

CONTENTS

ILLUSTRATIONS

FOREWORD

Patricia Austin Becker's *Cane River Bohemia* induces a palpable sense of time and place, along with a wistful yearning for the days when letter writing was the primary means of long-distance communication between friends. At the same time, she has created a fascinating, thoroughly researched study of the force of nature that was Carmelite Garrett Henry of Melrose Plantation at Isle Brevelle in southern Natchitoches Parish, Louisiana.

Accomplishing both in the same work is quite a feat. Becker educates us on the significant events of Cammie Henry's life, as a good biographer should, but at the same time captures the feel and taste of life on an isolated Louisiana cotton plantation and the "mystique" of Melrose, which, as the author says, "infuses the place like the mist off the Cane River."

"Miss Cammie" was a widow at forty-seven with eight children, attending to the mundane details of running a profitable agriculture enterprise while creating an artists' colony at Melrose that nourished a contingent of well-known writers, painters, and photographers.

The soul of this book is the relationship between Cammie and writer Lyle Saxon, naturalist and writer Caroline Dormon, and, to a lesser degree, writer Ada Jack Carver. Becker reveals their personal and professional interdependence through copious excerpts of letters between the four and provides the historic context in which these formal but highly personal, even intimate, letters were written. In fascinating detail, she describes how Cammie would forward letters she received to the others, so that the four of them would know what each was thinking. An inveterate collector of everything written, Cammie would insist that the forwarded letters be returned to her for inclusion in the myriad scrapbooks she kept during her life. These scrapbooks, along with the extensive contents of her library at Melrose, have been enshrined in the Cammie Garrett Henry Research Center at Northwestern State University of Louisiana in Natchitoches.

Never idle, Cammie Henry was a preservationist before it was fashionable, a generous and prodigious patron of all the arts, and a collector of documents, artifacts, and people. She held on to Melrose through the Depression, creating unique sources of income to support not only her family but also the many people on the plantation and in the community who depended on her.

The places and names that appear in this book—Melrose, Isle Brevelle, "The Normal," Briarwood, the Lemee House, Cloutierville, Irma Sompayrac, Gladys Breazeale, François Mignon, Clementine Hunter, the Roques, the Balthazars, and the other "Children of Strangers," to use Lyle Saxon's term, who lived down the river—are all familiar to anyone who lived in the second half of the twentieth century in Natchitoches, where I was born one year after Miss Cammie died at Melrose in 1948. Through the years I've been asked by many people in many places if I knew her.

My forebear, Londonderry-born Joseph Criswell Henry Sr., acquired Melrose in 1884, and his son John Hampton Henry, husband of Cammie, was the owner until his death in 1918 at the age of fifty-six. My cousin Pat is mentioned by Becker, described as a youngster playing at Melrose, climbing in the fig trees and building a set of stilts.

But you do not have to be related by blood or marriage to Cammie Henry or from Natchitoches or Louisiana to be enthralled with these people and places. The artists' colony Cammie Henry created at Melrose, at the intersection of three roads down the winding Cane River in cotton country, has had a lasting impact and will be appreciated by any sensitive reader.

The author has written an important book about a creative dynamo of a woman who was a force in the lives of her writers and artists, through whose work the world learned of the mystique of Melrose. This book reminds those of us who knew of Cammie Garrett Henry of her accomplishments and introduces her to others.

For that, and for the emotions I felt while reading, I am forever grateful to Pat Austin Becker.

—MICHAEL HENRY

PREFACE

One cold day in February 2013, I was at Melrose Plantation, standing in Cammie Henry's bedroom with a tour group, full of questions. It was certainly not my first tour through the plantation, but it was the first time I realized that there needed to be a record beyond what was currently available concerning her accomplishments. I was fascinated by the fact that this woman had opened her home to writers, artists, and photographers who wanted a quiet retreat in which to work. What motivated such a thing? What was the plantation like then, with Cammie and Lyle Saxon sitting on the gallery in the late afternoon, watching the sun set over the Cane River and talking into the night? Why hadn't her story been told?

As I began to search for answers, I found that there were two or three brief biographical sketches of Cammie's life as well as a couple of good dissertations. But the treasure trove, the real insight into who Cammie Henry was, came through her own words, in her letters and in her scrapbooks. There she really began to come alive, reminding me sometimes of women in Kate Chopin's novels yet vibrantly distinct in her own right.

What Cammie Henry accomplished in her lifetime is nothing short of remarkable, and she did it all with good humor and a positive, though sometimes stoic, outlook. Her nostalgic perspective and appreciation of the plantation way of life informed her entire adulthood as the "mistress of Melrose." Accomplished in weaving, quilting, and gardening, she managed to raise and educate her eight children and maintain a working cotton plantation through the Great Depression. She was an indefatigable preservationist. She acquired and moved old structures to her plantation and rescued, restored, and repurposed cast-off antiques from neighboring estates. She amassed over two hundred scrapbooks filled with newspaper clippings, letters, magazine articles, photographs, hand-written notes, and ephemera. Her personal correspondence with friends, pub-lishers, writers, librarians, artists, photographers, and researchers is staggering. Her collection of books and diaries, each one selected for a specific reason, in-cludes a large concentration of works by Louisiana writers and about southern history in general. Many of those books are rare, some are not, some have her personal notations in them, but they all help define the mistress of Melrose.

She was also a stalwart patroness of like-minded creative artists and scholars, offering Melrose as a haven where their output could flourish. Writer Lyle Saxon famously declared he could "write a book" there. Author Ada Jack Carver told Cammie, "The Land of the Lotus has meant more to me than I can ever make you know."[1] Writer Mary Belle McKellar described Melrose as "an Olympian fairy land of a plantation." New Orleans painter Alberta Kinsey came as often as she could, staying for weeks at a time, painting magnolias, cabins, and land-scapes. Writer, lecturer, artist, and conservationist Caroline Coroneos Dormon of nearby Saline visited frequently. Whether it was Cammie herself, the lush Cane River setting, or the association of fellow artists on-site, something about Melrose inspired its many visitors.

Cammie's relationship with her inner circle of friends goes to the very es-sence of her person: she was Saxon's muse, Carver's rock, and Dormon's best friend and biggest supporter. Her scrapbooks are filled with letters and notes from visitors thanking her for her generosity and hospitality. "She is the type of person," Saxon wrote, "who strikes you as being altogether alive."[2]

Kate Chopin was twenty years older than Cammie, but it's almost certain that they met in Natchitoches; they were related by marriage, and they had many common friends. Chopin's sister-in-law, Marie, was married to Phanor Breazeale, who was one of Cammie's nearest neighbors along Cane River as well as a bene-factor of the Natchitoches Art Colony with which Cammie was involved. Cora Henry, the sister of Cammie's husband, John, married Lamy Chopin, Oscar Cho-pin's brother. After Cora's death in 1892, Lamy Chopin married Fannie Hertzog of Magnolia Plantation, another neighbor of Cammie. By the time Cammie married John H. Henry in 1894, Kate Chopin's Louisiana husband had been dead over ten years, but she continued to visit her in-laws in the Cane River area, and it is quite possible that with so many common associations, she and Cammie crossed paths. Certainly, Cammie collected Chopin's novels for her library. Much later, Cammie's daughter, after marrying, would live in Chopin's Cloutierville home.

When I decided that Cammie Henry's story was one that I simply must write, I contacted Mary Linn Wernet, archivist at the Cammie Garrett Henry Research Center at Northwestern State University of Louisiana. Mary Linn was very clear about the scope of the collection, which is vast, but also very willing to help me. This book absolutely would not exist without her guidance and the quiet, patient aid of her assistant, Nolan Eller. Their hospitality, help, and good humor were priceless, and I'm forever grateful.

During the course of my research, I made many more visits to Melrose. Every time, I felt a sense of whatever mystique was there in the 1920s and '30s to inspire Saxon, Dormon, Carver, and Kinsey. Maybe it is the absolute quiet that descends at night, with only the sounds of the cicadas and bullfrogs singing under the soft glow of the stars and moon over the Cane River. Maybe it's something about the river itself; there is a local legend that once you drink the waters of the Cane, you are forever tied to it. This legend still exists: I sat in a local pub in Natchitoches one afternoon and heard the young lady behind the bar, who was a college student, tell a patron: "I don't even dip my toes in that river! One day I'm going to move from here, and I don't want to have to stay here just because I put my toes in that river!"

Whenever I walk the grounds of Melrose, I imagine what it looked like when Cammie Henry lived there, when Lyle and Ada sat on the swing on his gallery, talking about how to deal with sudden fame. I can hear the Henry children laughing and playing in the yard; I can hear the singing coming from the kitchen as meals were prepared or from the backyard, where the laundry was done. I can easily imagine the gaiety and camaraderie between Lyle and Cammie, the clink of ice in Lyle's highball glass, his aristocratic southern tones floating through the dusk as he read an intriguing passage from a novel or talked the latest plantation gossip. I can almost smell the strong, dark coffee as it was delivered each morning to the various cabins of the writers and artists in residence. I can see the chickens, the dogs, peacocks, Carrie Dormon's cats, Leudivine Garrett's quilts hung over the gallery railing, and Cammie herself in her galoshes and shirtwaist, watering the irises.

As I was doing the research for this biography of Cammie Henry, I found myself falling under her charms and almost felt a part of that Melrose circle myself.

Cammie was a vibrant, intelligent woman who was well aware of the changing times in which she lived. Opportunities for women improved significantly with advancements in education, the suffrage movement, and recognition in the workplace. She witnessed Reconstruction, Prohibition, and two world wars. While these events unfolded around her, Cammie both reached for and then turned away from the outside world to conduct her carefully orchestrated plantation life. She obtained her education, registered to vote, raised her children, and ran her plantation after the early death of her husband. Yet in some ways it was an anachronistic realm she inhabited, wearing long skirts and managing plantation staff as late as the 1940s.

Cammie was not unique in operating a rural plantation as a fairly young widow. This was the reality of many southern women immediately after the Civil War. What is remarkable about this southern woman of Cajun descent was the absolute determination with which she approached every task and the good humor she brought to anything she did. Her dreams and her grit together made for a fiery, independent-minded woman who rejected conventional proscriptions of what a proper lady should do. While other women were joining the club movement and attending meetings, Cammie was digging wild iris out of mud banks. "Who in the name of Heaven wants to be a perfect lady?" she once said.[3] In an age when most historians were male, Cammie documented the past the best way she could: through her voluminous scrapbook collection. She was an early preservationist who recognized that the South's traditional ways were quickly changing and that someone needed to record them. She chose to do that through her scrapbooks, book and manuscript collections, arts patronage, antiques salvaging, and textile crafts practices.

I have spent years reading her letters, deciphering her hurried scrawl, handling her scrapbooks, examining the records and documents she left behind, and experiencing vicariously the joys and sorrows of her life. I will leave it to scholars to write the critical interpretive study of Cammie Henry; my hope with this book is to take readers back to Melrose, both to the place where Cammie lived and to the way of life she nurtured. I hope to convey the sense of sitting in Lyle's cabin, reading before Cammie's fire, and laughing on the galleries of the big house with a group of friends. If it is episodic or scrapbookish in nature, then it at least reflects the medium by which Cammie preserved her own life.

If Cammie was my muse for this project, my family was my support, and I thank my husband, Steve, and son, John, for putting up with my long hours of research and writing. Margaret Lovecraft at Louisiana State University Press has been invaluable as she helped guide me to people and resources that could assist me. I am extraordinarily grateful for her encouragement and expertise. Tom Whitehead, with his vast bank of knowledge, was always helpful and quick to respond to any question I presented. I'd be remiss if I didn't thank J. Michael Kinney, who introduced me to the novels of Michael Henry in his Natchitoches bookshop one day. Michael Henry became a good friend, offering encouragement and penning this book's foreword, a gift for which I am eternally grateful.

Most of all, thank you to Cammie Henry, whose presence I felt every step of the way. I hope that this book helps keep Miss Cammie alive at Melrose to continue to inspire others.

A NOTE ON THE TEXT

Where I have quoted from letters or from marginal notes in the scrapbooks, I have left the spelling as is. Sometimes the name of Cammie's Melrose chauffeur is spelled *Fugaboo* and other times *Fuggerboo*. I have left these variations intact. Cammie often abbreviated words, such as *wkend* for *weekend* or used a symbol such as & for *and*. I have left these as well.

Cane River Bohemia

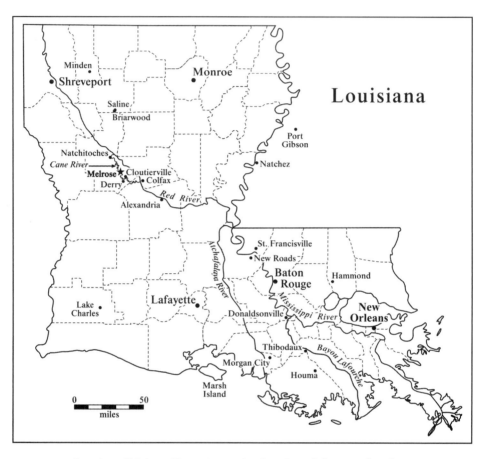

Location of Melrose Plantation on the Cane River below Natchitoches.

(Map by Mary Lee Eggart)

1

Finding Her Way, 1871–1920

Our gardens are fine and the poultry keeps me proper busy . . .
your father is planning to enlarge the Melrose store and do a big business.

—CAMMIE HENRY, letter to her son Stephen, March 30, 1911

In June 1920, after an absence of forty years, Cammie Garrett Henry returned to southern Louisiana to visit the plantation homes of her childhood. Though the house where she grew up had burned years earlier, on that visit in 1920 she took photos of old trees and the crumbling walls of the old plantation warehouse, noting, "The very brown earth brought memories!" Her eighty-year-old mother and her five-year-old daughter stood next to her as she snapped photos.[1] Born in 1871, during Reconstruction and only six years after the end of the Civil War, Cammie Garrett could not help but be influenced by the social upheaval that conflict caused and the subsequent attempt by southerners to reconcile their lives and values. She would remain loyal to the southern cause all of her life, but she was also forward thinking in her lifelong efforts to preserve Louisiana culture, architecture, and history.

Cammie was very much a product of her ancestry. She was the daughter of Leudivine Erwin and Captain Stephen R. Garrett. On the Erwin side of her family, Cammie's ancestors included Joseph and Lavinia Erwin. Joseph Erwin was born in North Carolina and fought in the Carolina militia during the American Revolution, rising to the rank of captain.[2] He married Lavinia Thompson, and they moved to Tennessee, where after much hard work carving out a life for themselves, Erwin established a plantation called Peach Blossom. It was in Tennessee where the famous duel in 1806 over a disputed horse race between a young Andrew Jackson and Erwin's son-in-law, Charles Dickinson, occurred in which Charles was killed. In one of her scrapbooks, Cammie notes that "the family never did forgive Jackson."[3] In 1807 Joseph Erwin, almost fifty years old, brought his wife to Louisiana from Tennessee on a flatboat down the Mississippi River. Their children stayed behind in Tennessee until Erwin was able to establish a home for them. He settled on a place on the Mississippi River about ninety miles north of New Orleans. He worked as a speculator and businessman, eventually making his fortune, and settled down to live the life of a planter on Home Place, one of his several cotton plantations that fronted the Mississippi for miles.[4] Joseph Erwin's Home Place was a beautiful and much-admired two-story residence furnished with mahogany, bird's-eye maple, cherry, and walnut pieces. Erwin lived prosperously for several years and settled his children on plantations of their own, giving them gifts of land and cash.[5]

Joseph Erwin's good fortune suffered a reversal in 1828, when the Mississippi River came through the levee and flooded the entire Plaquemine area. Crops were ruined, and Erwin sold one parcel of land after another. His debtors could not pay him, and he could not pay his own debts. His youngest son, Isaac Erwin, purchased his father's share of Grosse Tête Plantation, and on April 14, 1829, Joseph Erwin died, after which Lavinia undertook management of his estate. There is some discrepancy as to how Joseph Erwin died. In her thesis on Erwin, Alice Premble White says he committed suicide. Harnett Kane, the prolific writer on southern history, says that Joseph Erwin was found head first in a deep, wide water jar at the end of the porch one morning after a fretful night of pacing and worrying. Kane suggests Erwin may have slipped and fallen into the jug, which a servant had neglected to cover.[6] At any rate, under her careful management, Lavinia Erwin eventually settled her husband's debts and upon her own death left $121,313.67 to be divided among the heirs.[7]

Isaac Erwin was married two times: first to Mary Nichols of Tennessee and then to Carmelite Picou of Breaux Bridge, Louisiana. Isaac's first wife came with him from Tennessee, and they established their home on Bayou Gross Tête. They had five children, and Mary died in 1832. Isaac married again in 1838 to Carmelite, a fiery Cajun girl with whom he would have ten more children.[8] Such a large family required a large house, so Isaac Erwin built Shady Grove Plantation on a grand scale. It was a two-story brick mansion with towering Ionic columns and a narrow portico on the front. The two front rooms were each thirty feet square, and reportedly, it took one hundred yards of carpeting to cover the parlor. In the parlor were two huge mirrors that were imported from France and said to have cost over one thousand dollars each.[9] The ceilings on the first floor were twenty feet high, while the ceilings of the second floor were eighteen feet high.

The second daughter of Isaac and Carmelite was Leudivine Erwin, who was born in 1840 and later married Stephen Garrett at Shady Grove. Leudivine and Stephen's second daughter, Leu Carmelite—"Cammie"—was born on January 14, 1871. Thus, it was following some of the most fascinating times in the nation's history that Cammie Garrett, with her husband, John Henry, eventually established Melrose Plantation, a place that seemed to operate in a world of its own, where the plantation system still reigned, servants still delivered coffee to the rooms of guests each morning, hired hands still worked in the fields, and men tended to business while women supervised the household. Cammie learned to take great pride in her heritage and ancestral background.

Her ancestry was distinguished on her father's side as well. Stephen R. Garrett was born in Hot Springs, North Carolina, in 1838. During the Civil War he served as a captain in the Confederate army and in C. E. Fenner's Battery at Jackson, Mississippi.[10] He also served in Company B, First (Rightor's) Battalion, Louisiana Infantry, and as second lieutenant, in the Fourteenth Battalion (Austin's), Louisiana Infantry. These Confederate ties would define Cammie's perceptions of the South and seemed to lock her into a time period in history that she re-created at Melrose.

After the Civil War, Stephen Garrett settled his family along Bayou Lafourche in Assumption Parish, Louisiana, and worked for eight years as an overseer on Scattery Plantation, located about seven miles below Donaldsonville.[11] From 1811 until the early 1900s, steamboats traveled the bayou, which branched off the Mississippi River and cut through Louisiana's fertile sugarcane country. The owner of Scattery, Miles Taylor, was a native of Sarasota, New York, who had settled in Louisiana, where he was a congressional representative, lawyer, judge, and sugar planter.[12] Taylor, whom Cammie referred to as "Uncle Miles," was one of many lawyers who worked for the City of New Orleans on the famous Myra Clark Gaines case. Gaines contended that she was the legal heir of Daniel Clark, who owned a great deal of important real estate in New Orleans. The case dragged on for fifty-seven years and was one that Cammie followed; she clipped and pasted articles about the case in her scrapbook collection, writing a notation that Gaines finally won her case after Taylor died but that the city had paid him thirty thousand dollars a year for his work.

Stephen and Leudivine had two other children: Ann Steele Garrett, born December 7, 1868; and Stephen Nelson Garrett, born February 13, 1876. Ann died when she was eight years old of yellow fever and is buried in Thibodaux next to her parents; Stephen lived in southern Louisiana until his death in 1955. Life along the bayou for Cammie included studies, visiting nearby family and friends, and the occasional party or dance at a neighboring plantation. Her life-long practice of keeping scrapbooks began as a hobby when she was a young girl, when she clipped articles that interested her and reflected her avid interest in history and pasted them into scrapbooks. She was baptized at Shady Grove and spent many hours with relatives there. According to Henry E. Chambers in his *History of Louisiana,* Cammie graduated from the high school in Donaldsonville, although it is likely that in her primary years she received a private education from tutors at home, as was the custom for children on plantations at that time.[13]

As she grew into a young adult, Cammie grew hungry for more education. The rural, isolated life along the bayou could not hold her or satisfy her inquisitive mind, so she pleaded with her mother for permission to attend the Louisiana State Normal School in Natchitoches. Women's choices in a plantation society were limited by the expectation that they either marry well for the purpose of maintaining or improving their social standing or that they take an acceptable position—as a teacher, for example. The normal school in Natchitoches was founded in 1884 for the purpose of training young women to become school-teachers. Previously, women's education had been limited to simply teaching young ladies to run a proper home or to converse with their future husbands. The normal schools opening across the nation provided new opportunities for women. Training women for future employment was a progressive idea, and teaching school was an acceptable option for young Cammie, who certainly also wanted to continue her own education. The first normal school students in Natchitoches enrolled in 1885; Cammie signed her pledge with the school in October 1887, and she graduated in 1891.

Cammie's decision to enroll at Normal was likely due to a combination of factors. There is evidence that the family had some financial troubles. In 1884 and 1885 Leudivine and Stephen took out several loans on their home; the papers for these loans can be found preserved among Cammie's own papers. Two "notes taken out on our home" in January and in March 1885 signed by Leudivine and Stephen are tucked into a file along with a letter from a North Carolina relative expressing remorse over Stephen's drinking and concern that it would "put an end to his days."[14] If Cammie could further her education and gain respectable employment as a teacher, this could only help the family. She was eager to go, so Leudivine granted permission. Leudivine had attended Salem Academy in North Carolina and would have supported her daughter's desire for a higher education.

After her graduation, Cammie returned home to her parents and completed her agreement with Normal to work in the field of education, serving for two years as teacher and principal at the small, private Guion Academy in Thibodaux.[15] School started after Labor Day and usually went through July. Along with managing the daily affairs of the school, Cammie also attended principals' conventions and conferences throughout the year.[16]

While a student at Normal, Cammie had met John Hampton Henry of Natchitoches Parish.[17] John's father, Joseph Henry, from Londonderry, Ireland, was from a wealthy family and had been educated at private schools; his mother, Marie Ausite Roubieu, from Natchitoches,[18] was also from a family of considerable wealth. Joseph was quite popular in Natchitoches Parish, serving as a state senator from 1884 to 1892 as well as representing the parish at the 1879 constitutional convention. He was the largest individual taxpayer in Natchitoches Parish at one time and was generally considered a kind and generous person.[19] He died leaving vast holdings, including prime real estate in the town of Natchitoches itself. His obituary in the *Natchitoches Enterprise* described him as "a most lovable man," with "strong likes and dislikes," but one whose "friendship was something to be counted on."[20] Many years later Cammie would tell a story about her father-in-law: "It seems that one of his sons wanted to employ some one in France at a price of one thousand five hundred dollars to trace the Henry family tree, but old Joe Henry, on hearing of it, peeled off a five dollar bill, tossed it to his son, saying: 'Here, take this five dollars and run for County Sheriff. As soon as your opponents hear of your candidacy, they'll tell you more about your ancestry than a flock of genealogists could ever find out, and you'll be saving a lot of money to boot.'"[21]

John Henry, born in 1862, attended Catholic schools, graduated at the age of twenty from St. Louis University with honors, and then returned briefly to Natchitoches to help his father farm his several plantations before going to New Orleans to work for eighteen months with R. M. Walmsley & Company.[22] That experience gave John a practical business background with a cotton merchant, and he returned to Natchitoches by 1884, a few years before Cammie began attending Normal. On John's twenty-first birthday, his father presented him with a plantation of his own near Derry, Louisiana, in Natchitoches Parish. In 1884 Joseph Henry also bought what is now called Melrose Plantation, located on the Cane River seventeen miles south of Natchitoches, in the Isle Brevelle community.

In time Cammie and John crossed paths. She was a beautiful girl with long, thick, dark hair, which she usually wore coiled on the back of her head, and large, flashing blue eyes. She was also educated, independent, and full of spirit. John was handsome and came from a privileged background, and he wanted to marry her. Cammie completed her two-year teaching contract and finally married John Henry in January 1894. She was twenty-three years old, and he was thirty-one. The afternoon ceremony was held in Thibodaux at the Presbyterian church, where Cammie and John said their vows under a wedding arch of mosses, cedar, sweet olive, and japonica. The couple honeymooned in New Orleans and then returned to the plantation near Derry. John continued farming with his father, and Cammie began having children.[23] Like many married women in the days before modern contraceptive devices, Cammie spent the next twenty years bearing children.

Their first child was born in November 1894 and was named Stephen after Cammie's father, reflecting her close tie to him. Stephen Garrett died two years later, at the age of forty-five, leaving his widow, Leudivine, nearly penniless after the estate was settled. She then divided her time between daughter Cammie and son Stephen, who lived in New Roads, Louisiana.[24] Cammie attended her father's funeral in Thibodaux. It reflected his Confederate military service, with a large number of his fellow veterans in attendance, serving as pallbearers and delivering the eulogy. Garrett was buried in the cemetery at St. John Episcopal Church next to his daughter Ann.

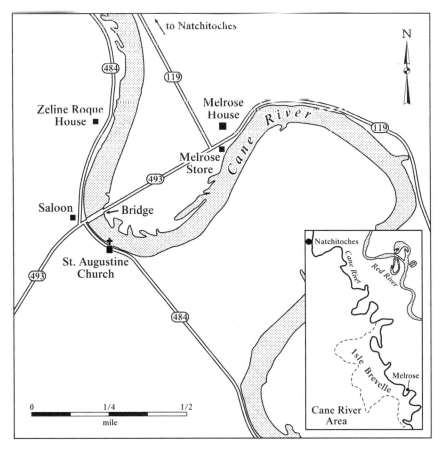

The Melrose area, ca. 1940, with highways superimposed for orientation.
Inset shows Isle Brevelle in relation to Natchitoches.

(Map by Mary Lee Eggart)

John Henry's father died in 1899 and left Melrose Plantation to John and Cammie. Melrose had been a favorite of Joseph Henry, and it was he who named the plantation after his love for Sir Walter Scott's poem "Melrose Abbey." Initially, Cammie wanted to stay in Derry. It was on the rail line, and she could receive mail two times a day there. Rural isolation did not appeal to her; she had left Bayou Lafourche to attend Normal for just that reason. John agreed to continue to live in Derry, but once Cammie realized that he would have to spend a great deal of time farming at Melrose, nearly twenty miles away, she insisted that they make their home there after all.

Melrose was located near the settlement of Isle Brevelle, a strip of land about three miles long between two rivers—the now-dammed Cane River to the east and the old Red River to the west. Isle Brevelle was home to a unique culture composed of generations of *gens de couleur libres,* or free people of color, blended from French, Spanish, Native American, and African heritage. Their descendants have a long heritage and tradition still celebrated today.

What is now called Melrose was originally a plantation established in 1796 by Louis Metoyer, a son of former slave Marie Thérèse Coincoin and Frenchman Claude Thomas Pierre Metoyer. When Pierre Metoyer arrived in Natchitoches in 1767, Coincoin was owned by the daughter of Natchitoches founder Louis Juchereau de St. Denis, and she caught Metoyer's eye. He was able to lease Coincoin, who then moved into his home. The two produced ten children, two of whom, Augustin and Louis, were important in the settling of Isle Brevelle. Because Coincoin was still legally owned by Marie des Neiges de St. Denis, her children were as well. Through the years Metoyer bought them and eventually purchased Marie Thérèse Coincoin; she bought the freedom of her five children born before her union with Metoyer. Coincoin and Metoyer parted in 1785, when Metoyer entered into a traditional, socially respectable marriage.[25]

In 1780 Metoyer had acquired a tract of land about eight miles below Natchitoches, in an area then known as "La Grand Côte" and now known as "Cedar Bend," along the Red River. He gave part of this land to Coincoin, and she later expanded her holdings through a petition to the Spanish government in 1787 and paid for with funds supplied by Metoyer. Later this section of the river would become the Cane River. Upon his death in 1815, Metoyer granted his children and Coincoin their freedom and gave them a parcel of land bordering his own. The property known today as Melrose was a tract granted to son Louis in 1796; on this Spanish concession in the early 1800s, Louis built the cabin known now as Yucca House on the east bank of his land grant.[26] In 1832 he began to build the manor house, or the "big house," but died before its completion. His son, Jean Baptiste Louis Metoyer, finished construction in 1833, setting the date in stone before the fireplace. The plantation stayed in the Metoyer family until 1847, when Théophile Louis Metoyer, young and inexperienced in his business dealings, lost the plantation.

It is a fascinating history and one that has been mired in confusion, reinvention, and revision for decades. The legend of Marie Thérèse Coincoin includes a series of tales and local lore that historian and author Elizabeth Shown Mills has called "historical chaos created by writers who theorize without adequate research."[27] It is not clear how much, if anything, Cammie Henry ever knew about Marie Thérèse; there is evidence, however, that in 1937 she was doing some research on previous owners of her home. She began to correspond with Cecile McClung in New Orleans, a relative of the Hertzog family, which owned Melrose in the years between the Metoyers and when Joseph Henry bought it. In a letter to an unidentified recipient, Cammie enclosed Cecile's letter and wrote, "Her g[rand]father was Hypolite Hertzog who owned Melrose—then Yucca—see if you can get a date near enough to do some research—I want to know when and how Hypolite Hertzog got Yucca."[28]

The origin of the Yucca name has been debated by scholars for many years, and this 1937 quotation indicates that Cammie believed her home to have once been called that. There is a letter written by the eldest son of Coincoin and Metoyer dated 1803 that is signed "Augustin Metoyer, Yucca Plantation."[29] Whether Augustin wrote that letter from his own home across the Cane River or from that of his brother Louis, the name Yucca was used when referring to various Metoyer homesteads on Isle Brevelle. Historians Gary and Elizabeth Shown Mills contend that the original structure on the property that would later be inhabited by Lyle Saxon was "dubbed 'Yucca House' by twentieth-century occupants."[30] The name of the fictional plantation in Saxon's novel, *Children of Strangers*, published in 1937, is Yucca. Authors Tom Whitehead and Art Shiver, in their biography of artist Clementine Hunter, credit the name Yucca to François Mignon, who visited Melrose in 1938 and came to live there in 1939.[31]

There was much work for Cammie to do when she and John moved to Melrose in 1899. The big house had been empty for several years and was almost uninhabitable. Broken windows, dirt daubers, spiders, cobwebs, and mildew infested every room. She set herself to cleaning and restoring the beautiful dwelling, ordering materials and drawing up plans.[32] It is believed that Séraphin Llorens of Isle Brevelle was the builder, and according to historian Gary Mills, his son

Hugh Llorens assisted in the remodeling of the home in the early 1900s.[33] Much of the renovation would likely have been done by hired hands already on the Melrose payroll and would not have been expensive labor. The work took place in the off-season, when hands were not needed in the fields or pecan orchards.

The big house was built with local cypress and brick and painted white. Constructed in the French-Spanish raised style typical of the area, it was one room deep and several wide. The lower floor often flooded when the river was high, before the Cane was dammed. The house went through multiple renovations, some during the Hertzog era and some during the Henry era. Today a wide gallery runs across the front with a gabled porch, and there is an exterior staircase at both the front and back of the house, allowing access to its second story. The railings are plain, functional, and non-ornamental. As their family continued to grow, Cammie added an octagonal garçonnière on each side of the house; the upper level provided extra bedrooms, and the lower level served as a screened-in summer dining room and additional living space. Shades over the screens could be raised or lowered as needed.

The ground floor is low ceilinged, with exposed timbers, and has a redbrick floor. The dining room has fireplaces at each end with nine-foot mantles. French doors open to the front and back galleries from every room, and plentiful windows allow for natural light. When doors and windows were opened, cooling breezes from the river, and the occasional bird, would flow through the house. As with most antebellum plantations, the kitchen was originally separate from the house due to fire risk. At Melrose it was connected to the house by a covered gallery, which was destroyed during the Civil War by Union forces and used for their fires. They camped at Melrose after burning nearby Magnolia Plantation. Cammie added a more practical service wing and later converted the old kitchen to a studio for visiting artists. The studio burned to the ground in 1965, and today only its large brick fireplace remains.

Over time Cammie filled her home with antiques and local pieces she found to be either interesting, historic, or beautiful. Many of them had come from the homes and cabins of her mulatto neighbors, as Cammie sought to preserve the culture of Isle Brevelle. It was not uncommon for her to negotiate for furniture or other items she liked. She would sometimes use vegetables from her gardens, groceries, or cash to barter.

Throughout the house were books. They were in every room and in later years were in the guest cabins as well. Cammie would spend the rest of her life at Melrose accumulating and reading books related to Louisiana, either by subject or by author, and books about southern history. In his 1941 book on plantation architecture, J. Frazer Smith writes, "On the interior [of Melrose] there is a space-giving atmosphere to the white plastered walls and woodwork, where books on top of books line the walls of the combination living room and library."[34] Furnishings included plenty of comfortable chairs and sofas through each room, with tables and reading lamps nearby.

By 1915 Cammie had given birth to seven sons and one daughter. Stephen Garrett Henry was followed by John Hampton, called "J.H.," in 1897; Isaac Erwin, called "Erwin," in 1899; Joseph Marion in 1902; Payne Walmsley in 1906; Daniel Scarborough in 1909; Robert Nelson, called "Bobbie," in 1912; and finally, Carmelite Garrett, called "Cammie Jr.," or "Sister" by her siblings, in 1915. Cammie spent her days raising her children, overseeing their activities and education, as well as managing her home, gardens, chickens, and other animals. In a letter in the spring of 1911 to eldest son Stephen, away at school, she wrote that the "poultry keeps me out of doors best part of the day!"[35] She also continued her own education, taking a series of correspondence classes through the Chautauqua Literary and Scientific Circle, a popular correspondence program. She studied Shakespeare and English history, among other subjects, and in typical fashion she saved her notebooks and papers. Cammie taught her children their elementary lessons daily from nine in the morning until three in the afternoon; classes were sometimes held on the back gallery of the house and eventually in a little schoolroom by the plantation store. In the tradition of plantations of the past, Cammie eventually hired two teachers to help educate her children as well as some of the neighborhood children who also attended classes there. She always left firm instructions that she was not to be disturbed when conducting school.[36]

Raising a house full of boys was never without mishap, and Cammie and John's sons all had typical scuffs, scrapes, and broken bones. In 1909 Stephen was injured in an accident with his horse and spent three months at Touro hospital in New Orleans. He was in a cast from his waist to his knees and had to rely on crutches for two months once he was able to return to Melrose. In 1912 six-year-old Payne had "a most unfortunate accident," which Cammie documented in his baby book: "Fell out of a hammock & ran a knife in his left eye—at first it did not seem serious—& he never lost the vision—was taken to Alex[andria] to Dr. R. F. Harrell. Eye was operated on & seemed to be a perfect success—when on the sec[ond] day it was discovered the good eye was sympathizing with the sore eye & fast going out. His beloved Bro. Stephen was with Payne thro[ugh] it all, night & day. The devotion was beautiful—devotion on both sides—On July 4 the little eye to be removed & done promptly to ward off utter blindness. His father & mother, Bro. Stephen were all with Payne—operation a success."[37] Payne was in the hospital in Alexandria for ten days; one month after the operation, he was fitted with an artificial eye.

There were also accomplishments to be celebrated. In 1911 J.H. won a Boys Club contest through the Louisiana State Fair for raising the most bushels of corn per acre. His prizes included a trip to Washington, D.C., a watch and fob, and various sums of cash from contest sponsors.[38] Older brother Stephen had won the same competition the previous two years.

When Cammie and John moved to Melrose, the land was bare. Former owners had farmed right up to the front doors, as was the custom. Cammie spent the rest of her life laying the foundation for and working in what would become for a time one of the most beautiful gardens in the state. It was work she enjoyed immensely and pursued with great enthusiasm. John spent his time managing the cotton and pecan plantation and the store. He often went into Natchitoches on business, a trip that at that time would take most of the day because of unpaved roads. John, along with his two brothers, inherited a great deal of land from their father, both in the city of Natchitoches and in the parish. The conveyance books record many land transactions between John and private individuals as well as civic parties such as the school board. He bought, sold, and leased land to various parties. He rented out the Derry plantation after the family moved to Melrose and may have sold the property, as it was low-lying land and not as suitable for farming as the Melrose property.

John wrote letters to his older sons, who were away at boarding school, briefing them on daily affairs of the plantation or teaching lessons of frugality. No personal letters from him to Cammie survive in her collection, but there are many letters he wrote to his boys. He often included a small check for pocket money or for some other designated purpose, and his tone was always affectionate.

John's side of the family reportedly had a history of kidney disease that was then referred to as "Bright's disease," and he suffered with it for some time. Bright's disease was often accompanied by hypertension, edema, and kidney stones. In 1917 John went to a clinic in Battle Creek, Michigan, for treatment. That June, Cammie's cousin in North Carolina wrote to say that she was sorry to hear of John's ill health and hoped "he will be benefitted at Battle Creek," noting that she had a friend there who was also seeking treatment for the same disease.[39] At the time, the clinic was considered the best facility in the country for treating Bright's disease. In the fall John returned home, feeling much better. The reprieve was only temporary, however.

In January 1918 son Joe, who was attending school at Chamberlain Hunt Academy in Port Gibson, Mississippi, wrote to his mother, asking: "How is father getting along now? I surely hope he is feeling better."[40] On the afternoon of Saturday, March 2, John was working as usual at the plantation store. The store stood a short distance from the main house, closer to the Cane River. It was a white clapboard, single-story building with a gallery across the front. "Jno H. Henry: General Merchandise" was lettered in black across the front. He was at his desk, talking business with associates, when he complained of not feeling well. He stood up and walked toward the door, and suddenly he collapsed. Acting quickly, his friends carried him out to the gallery of the store into the fresh air, where they tried unsuccessfully to revive him.[41] Dr. Keator, of Bermuda, attended the body and listed the cause of death as cerebral hemorrhage. Cammie, at forty-seven years of age, became a widow with eight children and a large plantation to manage.

John's remains were taken to Natchitoches the next day for burial in the American Cemetery. The service was conducted by Reverend J. Orson Miller, then rector of Trinity Episcopal Church in Natchitoches.[42] Stephen came home from nearby Camp Beauregard and J.H. from Louisiana State University to bury their father and support their mother. Erwin and Joe were not able to get home from boarding school in Mississippi in time for the funeral, but they arrived shortly thereafter to mourn with the family. In a March 17 letter, Erwin ex-

pressed to Stephen his disbelief at the sudden loss of their fifty-six-year-old father: "I can hardly realize father is dead . . . he was certainly good to all of us."[43]

By Thursday of the next week, John Henry's will, dated December 14, 1914,
was filed and read in Natchitoches. The handwritten document, drawn eight
months before Cammie Jr. was born, was found in his iron safe. In it he left
charge of his estate to Cammie and Stephen and wrote of his desire "that it will
never be God's will for my seven sons to have a stepfather, but if so, my oldest
son Stephen G. Henry and my dear wife Cammie Garrett Henry are to be sole
rulers over the younger children until they become of age, and no other ruler on
this beautiful and good earth." John also expressed his spiritual faith and said, "I
believe strongly in the Roman Catholic Church and do pray that some of my sons
will be Catholics."[44]

John's devout Catholic faith must have been a source of conflict in his marriage. That he and Cammie were married in a Presbyterian service is worth noting; even more interesting is that Cammie selected a Protestant burial for her
Catholic husband. His hope that his children would become Catholic did not
happen under her guidance. She steadfastly raised her children in the Protestant
faith. Cammie appears to have treated her husband with all due respect and propriety, and in her correspondence with her children, she always reminded them
to write to their father and to thank him if he sent them money while away at
school. She encouraged them to honor and obey him, but raising them in the
Catholic faith never seemed to be part of her plan, despite her husband's wishes.

There is little evidence of the nature of Cammie's relationship with her husband; it is interesting to note again that in the reams of correspondence that she
saved, there is no letter from him to her. There is nothing documenting their
courtship. There are no wedding photos, and there are no other photos of John—
around Melrose or with the children—in Cammie's collection. She pasted no
photos of him in the scrapbooks she made for her children. There is little doubt
that this absence is intentional; this is a woman who saved nearly every letter,
photo, envelope, and assorted ephemera that she ever touched.

There have been rumors that John had a volatile temper, and he was even
charged with assault and battery against the daughter of his business part-
ner J. M. Gorum, a charge of which he was cleared by the courts in 1915.[45] Regardless of the judicial outcome, it was an incident that lent weight to the rumors and certainly would have embarrassed Cammie. No doubt, Cammie was

stubborn and strong-willed in her own right, and if she won the battle over religious education, she probably won many other battles as well in their marriage. The nearly total absence of John Hampton Henry in her archives certainly suggests that.

After John's death, Cammie was in control of Melrose, the family finances, and her children. J.H. left college to help with the plantation because Stephen had his military career. J.H. also saw to practical matters, such as taking out insurance, a plan that Stephen fully endorsed, writing to his mother, "I am so glad you and John took out the insurance—all business people do this for everyone's protection."[46] It was also clear that they would need to engage "a first class bookkeeper" to help manage the business end of the plantation.

The same spring that John died, the Spanish flu broke out in the United States, believed to have started in an army camp in Haskell, Kansas, although there is no way to know for certain.[47] The first wave was relatively mild compared to the second and third waves that came later that year. As soldiers in overcrowded camps in this last year of World War I prepared to ship out overseas and move eastward, the flu moved with them. When they left camp and went into the civilian population, the flu went too. By the time the pandemic was over, in 1920, it would kill at least twenty-one million people worldwide. In his study of the outbreak, John M. Barry notes that what made this particular epidemic so devastating was that unlike a normal flu outbreak, which is usually most lethal to the elderly and infants, this one was especially lethal to men and women in their twenties and thirties.[48] The 1918 flu virus attacked its victims with astonishing speed.

Both Erwin, sometimes called "Ikie" by the family, and Joe were focused on their studies at Chamberlain Hunt in the fall of 1918, still mourning the loss of their father but certainly aware of the flu epidemic, as panic began to spread across the nation. In mid-October, Joe wrote to his mother to tell her of "a little riot" at the school as fifty-five boys were preparing to leave Chamberlain Hunt in the hope of avoiding exposure to the virus, although he notes that there had been no cases of the flu there at all so far. "This is about the healthiest place that I can think of," he said. Joe described a chaotic scene as some of the boys tried to run away from the school, and those who did not want to leave were called "cowards."[49] At that point forty boys had already left.

Two weeks later, Erwin wrote to Cammie expressing gratitude that nobody at Melrose had the flu; it is the only mention of the virus in this particular letter.

On the same day, Erwin also wrote to his grandmother Leudivine and told her that "not any of us has the influenza at C.H.A. yet."[50] Both letters are otherwise unremarkable and filled with his customary observations about the weather and assurances that both he and Joe were working hard on their studies. Within two weeks, nineteen-year-old Erwin would be dead from the flu.

By November 10, just one week after that letter, Erwin was back at Melrose and in much worse condition than anyone had realized. As Cammie nursed her ill son, she kept a heart-wrenching log on a lined tablet, noting his temperature hourly and all medications dispensed. Erwin's fever was spiking around 102 degrees, and he was coughing a great deal. Pneumonia developed quickly with this strain of the virus, and as Erwin's lungs filled with fluid, the coughing was almost constant. Cammie stayed by his bedside and sponged him with a cool rag. She gave him aspirin to reduce the fever. By November 12 he was vomiting, refusing medicine, and unable to keep any liquids down. The vomiting continued for four more agonizing days. Cammie documented his final hours:

> Broth at 12 (chicken broth). Dark urine.
> Beef broth at 3:40
> Temp 101
> Broth at 5:30
> Broth at 7
> Perspiring freely
> Death at 8:20. God still reigns.[51]

Just eight months after the loss of her husband, Cammie lost a son. For the second time that year, the Henry family traveled to the American Cemetery in Natchitoches, where Isaac Erwin Henry was buried near his father. Stephen was in Europe attending machine gun school as part of his service with the National Guard and did not learn of his brother's death until late January. His last letter to Erwin is dated November 17, 1918, in which he promises to be home soon, but Erwin would never receive that letter. Stephen was devastated that he could not help or console his mother over Erwin's death, and he ached to return to Melrose. It would have been of some comfort to Cammie that her mother began to stay full-time at Melrose after Erwin's death rather than commuting between Cammie's home and her brother Stephen's.

Family friend Dr. Milton Dunn, who was living at Melrose, wrote the obituary, in which he said: "This was one of the most precious hearts God ever created—so thoughtful and considerate for others. His character was superb. Gentle and loving in his disposition; devoted to his darling mother."[52] Dunn had been a friend of Cammie's father. The two men served together during the Civil War, after which Dunn returned to Grant Parish to practice medicine and serve as coroner. He has been referred to as the first writer in residence at Melrose and in 1920 published his "History of Natchitoches" in the *Louisiana Historical Quarterly.*

Condolences poured in. Cammie received notes and telegrams from business contacts, from both her own and Erwin's friends, and from family members scattered across the country. Her brother Stephen sent a letter noting that "it is the prayer of your brother that your Christian spirit will help you to bear this new sorrow."[53]

During Erwin's illness, Cammie wrote a letter to the administration of Chamberlain Hunt Academy apprising them of Erwin's grave condition; they received it on the day of his death, along with her subsequent wire notifying them that he had expired. The president of the school responded promptly, expressing great sadness, and he assured Cammie that Erwin had been sick less than twenty-four hours. Out of an abundance of precaution, he said, they kept him in bed.[54] In a December letter, he further explained that the flu had come quite suddenly to the school: "In three days we had more than twenty boys sick. We were forced to open rooms adjoining the hospital for the convenience of the doctor and of us who nursed. We placed twenty-four boys in the four rooms."[55]

Years later Cammie uncovered a letter she had sent to Erwin saying: "I'm afraid Spanish flu—going to get all over the country. Try not to take cold."[56] She wrote on the back of the envelope "Found in his old trunk 25 yrs after, 1943," and tucked the letter into a scrapbook for safekeeping.

Cammie kept busy responding to the condolences, but she continued to struggle with the tragedy for months, finally reaching out nearly a year later for spiritual consolation from a longtime friend from her days in Thibodaux, Reverend Edward Young. In his response he noted "the sad tone" of her letter and urged her to "not grieve so much for the departed one" for surely he was now "present with the Lord . . . Our duty now is to the living." He reminded Cammie that "God has greatly blessed you in giving you eight children, only one of whom

He has permitted thus far to pay the last debt to nature." Reverend Young urged Cammie to dry her tears and to "take up life's duties cheerfully and uncomplainingly for the sake of those whom He has still left you" and to continue ministering to the living. Taking a very personal note, Reverend Young also confessed to Cammie that he had always been concerned about "the wide difference in religion" between her and her husband. He complimented her decision to "remain steadfast" and to raise the children "as Protestants and Presbyterians." "You are wise," he continued, "in sending your boy to Davidson College where the influence is so strongly religious."[57]

In dealing with her grief by focusing on the living, Cammie came to a difficult decision regarding her youngest son. Bobbie had physical and learning disabilities and was developing slowly, which concerned his mother deeply and which she found increasingly unable to manage alone. He had been born without a hand, and Cammie was advised by doctors at the time against a prosthesis. She took him to doctors in Shreveport and around the state searching for help, and she hired a nurse to assist her at home in caring for him. In the scrapbook she made for him, Cammie wrote that "Bobbie had a hard baby hood—born afflicted—nothing normal for first 2 yrs . . . was 3 yrs old before he even walked alone."[58]

In June 1919 Cammie and Stephen made the decision to place Bobbie, now seven years old, in a private facility that could better handle his needs. It was an extremely hard choice for Cammie and came only after she had conducted a great deal of research. She read and collected countless articles, wrote to doctors, and researched institutions all over the country to find just the right place for her son. She eventually settled on the Bristol-Nelson School in Murfreesboro, Tennessee. Inside a booklet describing the school and its program, Cammie wrote: "Where Bobbie was put June 21, 1919. Strangers will be more patient with him than his own."[59] Heartbroken but stoic, Cammie did what she felt was best for her son's education and physical care.

The Bristol-Nelson School was located on five acres of land in the middle of Tennessee; the main building was a colonial brick mansion with fourteen rooms with French windows opening onto a shaded veranda. Cammie respected founder Cora Bristol Nelson (Mother Nelson) and her work with children, and they maintained a lifelong friendship and correspondence. In the margins of the informational booklet, alongside a paragraph that describes Bristol Nelson's work with special children, Cammie wrote, "Isn't she a wonderful woman!" The school was limited to fifteen resident students and took no day students, so each child was able to receive individual attention and care. It was important to Cammie that Bobbie be as near home as possible and that the school not be "too big."[60] Though schools for children like Bobbie were difficult to find at that time, especially ones that met her criteria, Cammie was adamant that she did not want him in an institution—she wanted to ensure he would get individual attention and care as well as an education in a loving atmosphere that was as much like home as possible.

After giving Bobbie time to settle in and get adjusted, Stephen went to Tennessee in October to visit him and wrote to his mother upon spending the day with him, "Bobbie is greatly improved."[61] The boy was glad to see his older brother and recognized him immediately. Stephen purchased some winter clothing for him and met his teachers.

In May 1920 Bobbie was scheduled for dental surgery to take care of an impacted baby tooth, so J.H., Cammie, Dan, and Cammie Jr. boarded a train and headed to Tennessee. Stephen stayed with Payne at Melrose, where they spent their days listening to the Victrola in the library, watering the garden, writing to Cammie, taking care of the animals, and tending to other household chores. Bobbie's dental surgery lasted an agonizing five hours while the doctors found, and then finally extracted, the offending tooth. Upon her return to Melrose, Cammie documented the trip in Bobbie's scrapbook: "Went to see Bobbie that night—7:30 as soon as we arrived. He recognized none of us! It was a dreadful shock at first. Then afterwards I realized it was another blessing that came to me in disguise. Had he remembered us & grieved for an absence how dreadful that would have been. How often we grieve for our own dear selves when we might imagine it is grief for the loved one!"[62]

They stayed for three days after the surgery to be sure all was well, and of their departure, Cammie wrote: "I believe he realized what parting was because the a.m. I left—his little head drooped—he never smiled or had anything to say—then I would have been glad had he not known me. Some things cut very deep into the human heart—and only a mother can appreciate some situations. The penalty for being a mother."[63]

In January 1921 Cammie Jr., now five years old, suffered a ruptured appendix and had to be taken to the hospital in Shreveport, thus requiring Cammie's absence from the plantation for a few weeks. Twelve-year-old Dan stayed
at Melrose with Dr. Dunn and wrote to his mother in Shreveport to reassure
her that they were busy as ever grinding corn at the gin and watching for rain.[64]
Stephen, who had mustered out of the service in 1919 and then reenlisted in
September 1920, wrote from Camp Pike, Arkansas, expressing gratitude that
his sister suffered "no bad after effects" from her illness, and he implored his
mother not to return home too soon, as "we live too far from an operating table
to leave Shreveport too early!"[65] The next month he sent a two hundred–dollar
check to Cammie "to use as you see fit" and promised to send her money each
month to help her with expenses, money she usually turned over to J.H. to use
in running the business end of the plantation.[66] Stephen's choice to reenter the
military could perhaps have been motivated in part by the desire to help his
family financially.

By the end of 1921, both Payne, age fifteen, and Dan were enrolled at Chamberlain Hunt Academy, following in the footsteps of their older brothers Joe and
Erwin. Dan seemed to be settling in and working hard at his studies; his grandmother Leudivine wrote letters of encouragement, telling Dan, "I knew you
would miss home and loved ones but no more than we do you, and you know there
is no chance for an education here—and a large school fits you to grapple with
this big old world when you have to strike out for yourself."[67] J.H. wrote in more
paternalistic tones; when Dan took it upon himself to drop his Latin class, J.H.
said, "You know Dan as well as I do, that this will not do." He directed both Dan
and the president of the school to add Latin back onto the schedule. J.H. reminded Dan, in emphatic capital letters, that "this FAMILY must EACH AND EVERY
ONE WORK AND TRY TO MAKE SOMETHING OUT OF THEMSELVES."[68] This was the
work ethic and sense of honor that Cammie Henry, and perhaps John Henry as
well, instilled in the children and one that they lived by for the rest of their lives.

Dan continued to work hard, but the school was not a good fit for Payne. He
was unhappy there and by December had left on his own for Baton Rouge, where
his brother Joe was attending Louisiana State University. Frustrated at Payne's re-
bellion, J.H. gave brief thought to driving Payne right back to Chamberlain Hunt
but decided against it. Instead, he wired Joe to put Payne on a train for home so
he could best determine how to solve the problem. Cammie was worried sick

about the whole affair and continued to insist that a good education remained of critical importance. She was endlessly grateful for the stern guidance of J.H. with the boys and wrote in her scrapbook, "As fine a son and [brother] as ever drew breath—not his equal in the land."[69] By early 1922 Payne was enrolled in public school in nearby Campti, much closer to Melrose, where he was much happier. With all of her boys now settled in school and growing into adulthood, Cammie continued into the 1920s focused on her young daughter and freer to pursue her own interests.

Though there were twenty-two years between Cammie's oldest and youngest children, she devoted herself to keeping them all close and the family bonds tight, and as the letters between them show, the children adored her. When the boys were away at school, they wrote letters to both Cammie and to Leudivine almost daily. In return Cammie sent news of home and boxes of items to them that might include anything from underwear to fried chicken. She sent money when needed but also taught the lessons of frugality and careful spending. There were also the occasional family spats and difficulties. She and eldest son Stephen had some sort of disagreement in the summer of 1923 that left her distraught and in an uncharacteristic depression for several months. There is a long, troubled letter that she wrote to him (but never sent) in which she said, "Part of me died last summer" and "[You] said right when you remarked I had lost the steel in my soul."[70] Whatever the trouble was, it was soon resolved, and the letters between them by the next month were once again loving and affectionate.

Despite the strong family bonds, Cammie and J.H.'s wife, Celeste, whom he married in 1924, did not always see eye to eye. There is a distraught letter from Stephen's wife, Mamie Gray, to Cammie in which she expressed deep hurt over remarks J.H. had accused her of making: "I haven't ever made the remarks I was accused of—Of course I can hardly expect you to believe me when J.H. has told you differently."[71] Stephen confirmed his wife's position in the misunderstanding when he wrote his mother, "M.G. never said or intimated that you were ever unkind to her—and for [J.H.] to quote her as saying such a thing was criminal . . . [it] is beyond my understanding."[72] Things were not much better in September 1924, when Celeste was in Touro Infirmary in New Orleans; Cammie wrote to her friend Carrie Dormon, "If she was speaking to me I'd go to her & bring things to a close."[73] But this too would pass, the family would come back together, and fences would be mended. In June 1925 Lyle Saxon, who had by then become a regular presence at Melrose, gleefully notified Carrie in a letter that "Celeste and Aunt Cammie have kissed (literally) and made up. Celeste came over and made the apology in quite the Grand Manner, I believe . . . and stayed for supper!"[74]

Through examination of her almost daily letters with her children while they were at school, it appears that Cammie Henry was as fine and loving a mother as any child could want. She may not have been physically demonstrative, though. In at least one interview, her grandson Stephen Jr. was quoted as saying that Cammie was "a very austere person and didn't enjoy showing affection for the members of her family," adding that he could not remember ever being hugged by her.[75] It is possible that Cammie saw open displays of affection as undignified, and perhaps this affected her relationships with her children and her husband. She clearly loved her children and later her grandchildren, expressing her affection in her letters. She managed to keep her family close, moderate disputes, and organize frequent gatherings at Melrose, especially the Sunday noon meal, which was always family time.

A good education for her children was of critical importance to Cammie, as that was what she felt was necessary to open the doors of success for them. Schooling options in rural northern Louisiana were not optimal then, so there was never any choice in her mind but to send the children away to learn discipline and obtain a well-rounded education. If military training and religious instruction were part of the package, all the better. In the scrapbook she kept for her son Payne, she wrote, "A boy needs a boy's school, boy's association, boy's military discipline, which I could not give you at home."[76] As the daughter and granddaughter of Confederate and Revolutionary soldiers, she was proud of her heritage and sought to instill that pride in her children. Stephen served in both World War I and II, eventually retiring from military service as a major general. J.H. successfully ran the plantation until his death in 1969, and her other sons worked in the private sector. Her only daughter graduated from college, married, and eventually moved to Shreveport, where she had children of her own. Cammie raised her children to be successful, thriving adults, but they were only the beginning of her legacy.

2

Congenial Souls Come Together, 1920–1924

Happy the wanderer who has, perchance, a friend who knows the mistress of Melrose;
who has indeed, a purpose; for this is the password opening the gates into
the lovely sunlit garden and into Miss Cammie's heart.

—MARY BELLE MCKELLAR, "Louisiana's Hardest Working Historian"

As the 1920s began, Cammie Henry was coming into her own; almost fifty years old, she had been a widow for two years, and her children ranged from toddler to adult—Cammie Jr. was four and Stephen twenty-five. No longer bound by the expectations of a husband, she was free to pursue her own interests, and she spent her time expanding the gardens of Melrose right down to the Cane River, working in her scrapbooks, conducting volumes of correspondence, and restoring buildings on her property that would become housing for writers and artists. At any given time, she was "Aunt Cammie," "Miss Henry," "Madame Henry," "Cam," or "Mother," depending upon what interaction was taking place and with whom.

Around the same time, the American South found itself experiencing a new literary and artistic renaissance, which was sparked largely by H. L. Mencken's scathing indictment of southern arts and letters in his 1917 essay "The Sahara of the Bozart." Prior to this literary renaissance, many southern writers were concerned primarily with romanticizing the antebellum South, writing plantation novels, and advocating the "Lost Cause" of the Civil War. Writers like George Washington Cable and Grace King examined both the pros and cons of southern culture and customs, while Kate Chopin's local color stories portrayed conflicted characters in strictly southern settings. Cable, for example, criticized racial inequities in southern race and class, while Grace King's celebrated "Monsieur Motte" told the story of a loyal freed slave.[1] Mencken dismissed the South as a wasteland of "worn-out farms, shoddy cities and paralyzed cerebrums." He declared that "the Civil War stamped out every last bearer of the torch, and left only a mob of peasants." Furthermore, he said, there simply are no art galleries, orchestras, operas, or public monuments worth looking at in this vast cultural desert. Mencken's essay produced a furious reaction in the South, and every warm-blooded southerner with an ounce of creative spark set out to prove him wrong. This emotional response resulted in an artistic renaissance that lasted at least five decades and has continued to evolve.

It is true, as John Bradbury points out in his book *Renaissance in the South*, that the "social upheaval" that followed the Civil War left many in the South with new perceptions and with "personal values" that "had been shaken, and their home towns, farms, and jobs had been largely transformed."[2] But as the literary renaissance developed, one group in particular, the conservative "Southern Agrarians," looked backward rather than forward; they often wrote about the romanticized past and promoted the old ways of manners and gentility over the new, industrialized, capitalistic ways even as they knew that those glory days were gone. This group of twelve was led by John Crowe Ransom, and they published their manifesto, *I'll Take My Stand*, in 1930.

The Agrarians were not unlike the American expatriates Ernest Hemingway and F. Scott Fitzgerald, among others, who after World War I were unable to reconcile themselves to the new social order and thus fled to Europe while at home new centers of literary activity were forming. In New Orleans the literary magazine the *Double Dealer* (1921–26) and at Vanderbilt University the *Fugitive* (1922–25) appeared. Some writers who subscribed to the Agrarian doctrine—such as Roark Bradford, Gwen Bristow, Harnett Kane, Frances Parkinson Keyes, and of course Lyle Saxon—were drawn to Cammie Henry and her Melrose Plantation, which grew into a haven for southern artistic creativity. Each visitor was attracted to Melrose for his or her own reasons. After Lyle Saxon took up residence at Melrose, some came because Saxon was there; as the years went by, others came because they had read about Melrose in their local newspapers, and some came simply for the atmosphere. The only common factors were the plantation and Cammie Henry.

In many ways Cammie never left the agrarian world that she was born into and in which she lived as an adult. Residing on a cotton and pecan plantation in the rural South, she retained many practices that harked back to days gone by, including keeping a house full of servants, some of whom lived in cabins around the Melrose property, and having coffee delivered by them each morning to the rooms of her guests. Even her mode of dress—a habit of white cotton shirtwaist and long black skirt, worn up until her death in 1948—reflected a more mannered way of life decades removed from the modern times in which she actually lived. Cammie tended to date things relative to the Civil War. For example, she might say that a quilt was "made before the Civil War," and in describing Shady Grove Plantation in her scrapbook, she noted that it was "built long before the War between the States."[3] She once told Caroline Dormon, "You know, I am not like other folks."

In their essay on this notable Louisiana woman, Lucy Gutman and Shannon Frystak describe Cammie Henry as "the living embodiment of the Lost Cause movement."[4] It would have been impossible for her to escape the influence of Lost Cause beliefs during her formative years. Cammie grew up hearing stories from her parents, grandparents, and other relatives about the South before, during, and after the Civil War and was familiar with the hardships of Reconstruction. She clearly admired stately antebellum mansions, noting in her childhood scrapbooks that she would love to have an old plantation home someday. But it is also safe to assume, based on a study of her life, that Cammie expanded her beliefs beyond nostalgia for the past or blind subscription to a political ideology. The Arts and Crafts movement, a growing national preservation movement, along with the many people she met and hosted and the wide reading she did, also shaped her. Her decades-long scrapbook hobby grew to include far more than pictures of pretty plantation houses. Her manuscript collection; her preservation of decaying cabins, furniture, buildings, and architectural elements; and her attempts to collect and record local family histories of her Natchitoches and Isle Brevelle neighbors as well as those of the servants who worked for her all offer evidence that there was much more to Cammie Henry's southern identity than adherence to the Lost Cause movement. Indeed, Gutman and Frystak note that Cammie "made great cultural contributions to Louisiana history" through her promotion of writers and artists who came to stay and work at Melrose.[5]

Gutman and Frystak also characterize Cammie as someone who refused "to abandon the elitist and racial assumptions inherited from her ancestors." As an example, they cite Cammie's display in her garden of a cannon used in the Colfax Massacre of 1873 as "an overt nonverbal statement to the hundreds of black house servants, field hands, and tenant farmers about the importance of knowing one's place."[6] A more thorough examination of Cammie's relationships and actions as the proprietor of a working plantation calls into question the motive Gutman and Frystak attribute to Cammie. While she was proud of her forebears' military service, it is doubtful that the cannon was ever intended as a subliminal statement to or a means of intimidating her hired staff or other people of color around Melrose. Every indication is that she was very kind and generous toward those who worked for her as well as her neighbors and that they returned her affection. There is just as likely to be an identification of Melrose employee Puny Count and his family tree in her scrapbooks as there is of one of the more influential and aristocratic families Cammie knew.

The Colfax cannon came to Melrose in 1921 through Dr. Milton Dunn, who had an avid interest in Confederate history. He had accumulated a large library of volumes related to the subject as well as a vast collection of correspondence with the soldiers he had known during the war and whose histories he sought. Upon his death in 1924, Dunn left his books and collection of histories to Cammie. Her primary interest in the cannon appeared to be preservation, similar to her dedication to maintaining rare books, deteriorating buildings, architectural pieces, oil paintings, and the lost arts of weaving and bookbinding. In Scrapbook 79 there is a photograph of the Colfax cannon in the garden at Melrose, its short, stubby barrel pointing upward. In the margin Cammie wrote: "Cannon taken from Capt. Boardman's boat. Easter time. April 13, 1873. Colfax Riot." She also indicated that the cannon was used in the Battle of Monett's Ferry in April 1864, which was a Union victory during the Red River Campaign of the Civil War. When the Henry family sold Melrose in 1970, the cannon went for $575 to a man from Colfax, and it is likely still in Colfax.[7]

As Gutman and Frystak point out, both Cammie and her daughter were members of the United Daughters of the Confederacy (UDC). Cammie's membership dates to 1920, and Cammie Jr. joined in 1956. There is no evidence that Cammie was ever particularly involved in the organization. The UDC was very active in the early 1900s, working to erect Confederate monuments in cities throughout the South to memorialize the sacrifices of family members. In her scrapbooks Cammie pasted clippings about many of these monuments and their dedication ceremonies. It is not surprising that she would be a member of this heritage group given her father's military service in the Confederate army, but she was not a very active one. She never enjoyed club meetings of any kind and seemed more interested in preserving clippings and photographs of the monuments than in attending their dedications.

Cammie was in fact among the early historical preservationists in the South, as her vast collection of scrapbooks attests. From childhood she saved newspaper clippings on subjects ranging from plantation homes and historic buildings to recipes and literary figures. The subjects covered in her scrapbooks expanded as she matured, but the intent was always that of historical documentation. She would have been aware that the federal government was working toward preserving American sites and that Congress established four major Civil War battlefields as national military parks between 1890 and 1899.[8] Preservation is something she had in common with Lyle Saxon, who gained a great deal of positive attention with his *New Orleans Times-Picayune* feature covering the fire that destroyed the French Opera House in New Orleans in 1919, an article found in one of Cammie's scrapbooks. Another early preservationist, Dr. William A. R. Goodwin, was instrumental in the restoration of Colonial Williamsburg in the 1920s. He noted "the spirit of ruthless innovation which threatens to rob the city of its distinction and charm," a remark that Cammie Henry could easily have made herself at various points in her life as she witnessed the encroachment of modern life in the form of automobiles, paved roads, and bridges on Isle Brevelle and later in Natchez, Mississippi.[9] In 1940 Melrose and other sites on Isle Brevelle would be photographed as part of the federal government's Historic American Buildings Survey. The HABS project began in 1933 and was an attempt to preserve a way of life and architectural relics that were quickly vanishing.

Cammie's attachment to the manners and customs of the Old South can be seen as her attempt to resist modern America and to preserve what she saw as a better, more refined way of living. Throughout her years on Cane River, she collected family histories and cherished antiques. In her papers can be found a vast collection of old quilt patterns, and she was known to travel the rural Louisiana countryside in search of local weaving patterns. She had looms and spinning wheels in multiple locations around her home and outbuildings and eventually devoted an entire cabin to looms. In a period when industrialization and modernization were growing, when people would soon travel by automobile on concrete roads, when personal items like clothing and books were made by machine, she was holding fast to the old ways. She saw value in thrift and hard work and expected others to do the same.

In the spring of 1920, Ellsworth Woodward, who was the first director of the Newcomb Art School in New Orleans, came to Natchitoches to paint and to lecture at the Louisiana State Normal College. Woodward was born in Massachusetts but followed his brother William, who was teaching art at Tulane, to Louisiana. Both Woodward brothers were strongly motivated to bring progressive ideas and artistic rejuvenation to the South. Perhaps influenced by landscape schools such as the Hudson River School, they saw great potential for distinctive southern art, and Ellsworth was so impressed with the beauty of the Cane River region that he returned in the summer of 1920 at the invitation of Ida Stephens Williams. During this trip he paid a visit to Melrose, where he painted a watercolor of the plantation. Woodward was a proponent of the Arts and Crafts movement, which began in Europe in the 1880s and spread to the United States as a reaction against modernization and machine-produced decorative arts. Richard Megraw describes the movement as one that attempted to reconcile "art and industry through tastefully designed objects of everyday use."[10] Ellsworth Woodward believed strongly in the aesthetic value of things, and he believed strongly that artists worked best in their own natural environment, so when his former students Irma DeBlieux Sompayrac and Gladys Breazeale, both from prominent Natchitoches families, decided to found the Natchitoches Art Colony, Woodward agreed to serve as instructor and critic its first year. He explained, "It is the duty of the Southern artist to depict on his canvas that indefinable something which makes the South a distinctive country."[11] The same dearth of creativity Mencken saw in southern literature, Woodward also saw in southern art.

While there were other landscape en plein air schools of art that had previously coalesced around the country, such as the Hudson River School, the Natchitoches Art Colony seems to have been rather insular in comparison. Its existence seems to have been nothing more profound than that Ellsworth Woodward found the Cane River landscape beautiful and inspiring. When Irma Sompayrac graduated from Newcomb and wanted to establish a vocation for herself in the arts, a landscape school seems to have been the perfect answer. Woodward was unable to commit himself as a leader of the school due to his already busy schedule, but one evening over conversation with Irma and his patron Ida Stephens Williams, during which they discussed the natural beauty of the region, he agreed to lend his influence and serve as instructor. Woodward convinced Irma that her local society connections would serve her well as leader of the art colony. Later he wrote, "Mrs. Williams' generous appreciation and real devotion to the beauty of that favored section of country was really the beginning of interest that was to center on Natchitoches."[12]

The first session of the art colony began on July 1, 1921; participants in the two-week session paid a tuition of fifteen dollars and were housed either in local homes or in the hotel in Natchitoches. All painting during the session was done en plein air. Lyle Saxon wrote about this first session for the *Times-Picayune* in 1921, two years before he ever met Cammie Henry.[13] As a benefactor of this new art colony and a friend to both Irma Sompayrac and Gladys Breazeale, Cammie pledged her support and volunteered her home as a place for the artists to paint. Each session of the art colony closed with an exhibition of the works, with cash prizes for the winners, and the first year, local planter Phanor Breazeale, Gladys's father, put up the prizes: one for watercolor and one for oil. There were about a dozen young women who participated in that first session.[14] Enrollment increased over the next few years, with sixteen students the second year, and the art colony was deemed a success, as the artists painted along Cane River, at Grand Ecore on the Red River, and at Melrose. A steamboat, the *Pearl*, carried the artists up and down the river to various scenic spots to paint. Caroline Dormon was a participant the second year, when Will Stevens, an art instructor at the Newcomb School of Art, served as instructor.[15]

By 1925 a rustic log cabin had been built on the bank of the Cane River at the north end of town to house the Natchitoches Art Colony and serve as a studio. Phanor Breazeale leased the little portion of land to the Art Colony, and Cammie dedicated many afternoons there planting the beds around the cabin with juniper, figs, willows, wisteria, and butterfly bush. Inside the little cabin was a bed for visiting artists that had a mattress made at Melrose stuffed with dried moss.[16] Along the walls were paintings by participants of art colony sessions, and there was a large window where one could watch the river course by.

The Natchitoches Art Colony preceded the less formal literary and arts colony at Melrose by a year or two, yet the two were closely intertwined. The Natchitoches Art Colony brought the artists to Melrose and the Cane River region to work, and Cammie's deep involvement as a hostess and benefactor perhaps contributed to her inspiration to have writers and artists stay at her plantation home. Cammie was not particularly social and did not enjoy the ladies' clubs and teas that were popular at the time, but she thrived on learning new things, and the art colony enabled her to bring interesting people with influence and information from other realms to her own backyard. This led to many of the friendships and associations that she would have for the rest of her life.

Caroline Dormon, known as Carrie to her friends, lived some forty miles to the north of Melrose in Saline, in Bienville Parish. She and her younger sister, Virginia, resided at their childhood home, Briarwood, having moved there permanently in 1917 after growing up in Arcadia, Louisiana, with their parents. Carrie's father was an attorney, and her mother had published a novel entitled *Under the Magnolias*, set just after the Civil War. Caroline's mother died in 1907, the year Carrie graduated from college, and her father died in 1909. Virginia, described as a raven-haired beauty, was "the indoor Dormon" and as likely to be reading Shakespeare or Milton as Carrie was to be painting flowers or studying birds.[17] Their cabin was built with timber from the nearby woods and surrounded by forest trails, a wild iris bog filled with pitcher plants, irises, and countless other specimens that Carrie cultivated and studied.

By August 1920 Carrie and Cammie were corresponding about Carrie's dear Kisatchie Wold, a project and a passion that would later define her legacy. There were almost seventeen years between them, but the two women likely met through their common interest in gardening and cultivating native Louisiana plants, shrubs, and trees.[18] They became friends for life and corresponded for the next twenty-eight years. Carrie was a tall, thin redhead, later described by David Snell, the son of Ada Jack Carver, as "all whipcord and piano wire." Snell's profile of Carrie Dormon in the February 1972 *Smithsonian* magazine depicts a woman more interested in studying leaves and squirrels outdoors than sitting inside burdened with domestic duties. Characterizations of Carrie as various trees or as the embodiment of nature itself abound through the letters and correspondence with her friends. Ada Jack Carver, upon sending Carrie a belt to a green dress she had given her, said, "You looked like a young willow tree in it!"[19] David Snell wrote that she would be as like to have "a cocklebur or two caught on the hem of her skirt" as not and that her "eyes were the green of chlorophyll." Lillian Trichel, a close friend of both Cammie and Carrie, said that "knowing Caroline is like knowing a lovely tree, fragrant and alluring, but always preoccupied with its own business of growing and blooming, always a little aloof from human problems."[20]

While Cammie had been collecting Carrie's published articles and pasting them into her scrapbooks, Carrie was exploring the forest of her childhood and taking pictures, some of which she sent to Cammie at her request. In mid-August 1920 Cammie wrote to Carrie thanking her for the pictures of Kisatchie, saying they were "just exactly what I wanted most. I've looked and looked at the pictures—they appeal to me. We must tramp every inch of that historic ground."[21]

As it would for others, Melrose offered a respite for Carrie from her busy schedule as a writer and a public school teacher and also provided a place where she could paint and work out stories. On one 1921 visit, she good-naturedly wrote out her own "Daily Program of Caroline Dormon at Melrose" and hung it up on the wall.[22] The schedule accounted for each hour of the day and had Carrie rising at 7:30 a.m. ("maybe"), writing, painting, fishing, talking, and going to bed at 9:00 p.m. Cammie often encouraged Carrie to quit teaching and focus on her writing and painting, saying, "You can't tell what you can do till you do it—& you can't do it [and] keep present job—that's one thing sure—burn the ships behind you & come stay in Melrose cabin a year—. . . you can write & paint—if you only let yourself—. . . can't serve two makers!"[23] It is easy to see why Cammie wanted Carrie to be free to paint; Carrie's watercolors of wild irises are luminous and stunningly beautiful. Carrie would always return to Melrose, but she never stayed an entire year as Cammie requested. She just never had that much free time.

In August 1921 Carrie wrote a history of Kisatchie at Cammie's request and sent it to her for the scrapbook. She began, "You have asked me for the story of our Kisatchie Wold Park, so here it is, to date."[24] In this three-page typed history, Dormon described her lifetime love of the forest and trees, which she thought of as her friends, and she detailed the steps to preservation she had taken to date: "From babyhood I have known and loved the long leaf pine forests of Louisiana. The happiest times of my childhood were spent in the long leaf pine country. These majestic trees have been inspiration when I was discouraged, my refuge in time of youthful sorrow, and silent witnesses of my childhood pleasures . . . When their devastation began it was almost more than I could bear."[25] She went on to explain that "after years of pondering and useless grieving for the pines," she had begun contacting forestry officials and went to the Southern Forestry Congress to solicit help. She finally decided that the best way to achieve her goal of preserving a section of longleaf pines for the future would be to approach the lumbermen themselves, a plan that eventually was successful.

Carrie made her living teaching school in those days. She taught both elementary and high school students through the years, beginning in Bienville Parish and later at the Kisatchie School, but she was deeply involved in the preservation of the trees and responsible for the eventual creation of Kisatchie National Forest. In 1919, according to biographer Fran Holman Johnson, Carrie wrote to the Louisiana Department of Conservation to inquire about the Southern Forestry Congress of 1920 because she was eager to launch her mission to preserve virgin tracts of longleaf pine forest.[26] In March 1920 she was appointed to the legislative committee of the Louisiana Forestry Association and was the head of the Forestry Division of the Louisiana Federation of Women's Clubs. In 1922 she attended a meeting of the Southern Forestry Congress in Jackson, Mississippi, where she met W. B. Greely of the U.S. Forest Service. She discussed with him her hopes of establishing a national forest in Louisiana.[27] She was also busy writing articles, which supplemented her meager income; one of her first to be published was "Highways plus Trees" for *Holland's Magazine* in 1923, about the need for intentional beautification of Louisiana roadsides.

Carrie spent most of 1923 employed by the state forestry department and traveled around the country giving talks on her work with young schoolchildren. The summer found her at the Georgia Forestry Association convention, where she spoke of the fascination young children have with the forest and their eagerness to learn about aging trees by counting rings, how to identify native plants and trees, and the effects of the forest soils and the water supply. Bessie Shaw Stafford, writing for the *Atlanta Constitution,* called Carrie "a very brilliant woman," which Carrie told Cammie was "very flattering."[28] By the fall of 1923, Carrie was tired of the bureaucracy of her job and had decided to resign her position effective in September.[29] While the loss of her job was a blow to her income, it did free her up to do some writing and painting and to spend time at Melrose. There was always the possibility that she might sell some of her work and the potential for literary success, although Carrie was not celebrity driven. What she craved most was perhaps simply financial stability rather than fame.

Carrie's passion for native Louisiana plants was evident in nearly every letter she sent to Cammie. In one she told of spotting some "cucumber trees" (a type of wild magnolia) from her seat on the train during a station stop near Packton, Louisiana: "My train stopped there 8 minutes—of course I had already spied the trees—blooming—so I soon had little boys scurrying off to get me a blossom. One came with a bloom just as train was pulling out. When I walked in the coach with it, everyone *gasped.* Nobody knew what it was. The flower measured 17 inches from tip to tip, and the leaves, which were not mature, 25 inches in length. The most tropical looking, startlingly beautiful thing I ever saw in La."[30] Carrie always carried a small spade under the seat of her car, and it was not un-heard of for her to pull over and carefully dig up a specimen of something she wanted to cultivate and study at Briarwood. Though the establishment of the Kisatchie National Forest was her greatest contribution, she also wrote poems and many short stories that she shared with friends, and she published many articles about Louisiana flora and fauna in various magazines and newspapers. She had studied literature and art at Judson College, in Marion, Alabama, and one of her friends there, Natchitoches native Ada Jack Carver, she would soon reencounter at Melrose.

Perhaps one of the most important people to enter Cammie's life was Lyle Chambers Saxon. Born in Bellingham, Washington, on September 4, 1891, Saxon was raised in Baton Rouge by his mother and her family after his father abandoned them shortly after Lyle's birth. Lyle always presented himself as a cultured, plantation-born Louisiana native, but if he was actually raised on a family plantation, it was probably a small place and perhaps even a combination of family places that he exaggerated upon in his imagination. Both his uncle and his cousins had small plantations where Lyle may have spent time as a child. He entered Louisiana State University in 1907 but, according to biographer Chance Harvey, did not graduate, leaving school with only three semester hours left toward his degree. Harvey speculates that Lyle left school in order to help support his mother financially.[31]

Lyle began working as a journalist in New Orleans in 1918, writing first for the *Item* and later the *Times-Picayune*.[32] Cammie and Lyle met in New Orleans in 1923 at the home of writer and historian Grace King during one of her famous literary salons. The southern literary renaissance inspired by Mencken was well under way by then, and King perhaps anticipated the philosophy of the Southern Agrarians when she told Lyle in an interview that she began to write from "a sort of patriotism—a feeling of loyalty to the South." She explained that the writers of her generation were "anxious to defend the South we loved when it was represented so badly in literature."[33] Cammie and King had exchanged correspondence both prior to and after the 1923 salon, and Cammie clearly admired *la Grande Dame* of New Orleans, as they both shared the same ideals about the Old South and about preservation, as did Lyle. The instant rapport between Lyle Saxon and Cammie Henry was destiny.

Lyle is consistently described as a charismatic and engaging individual, a man to whom people simply gravitated. Future Melrose resident François Mignon wrote in 1940, "For me, Lyle has always personified the Planter—inordinately tall, substantial in bearing, and sufficiently heavy to give added weight to the dignity of his carriage."[34] Biographer Chance Harvey says he had "classic good looks," with black hair and pale blue eyes; he spoke with a "soft, cultured speech," and he would often teasingly raise one eyebrow when he talked to you. Cammie Henry had followed Lyle's newspaper work for years before they met, evidenced by the clippings of his columns in her scrapbooks that predate 1923.

One can imagine Lyle's anticipation as this new southern renaissance exploded around him; he began the 1920s as a newspaperman, as a journalist, but he always wanted to write fiction, especially the kind that depicted his perceptions of the Old South and plantation life. Even though his one novel would not appear until the 1930s, Lyle's body of nonfiction revealed a certain southern urgency, a need to reestablish the validity of the agrarian South, and he must have felt the challenge to produce some fiction of literary merit that would ensure his place in this new movement alongside his contemporaries William Faulkner, Sherwood Anderson, Roark Bradford, and Hamilton Basso.

Cammie extended an invitation to Lyle to visit her plantation, and he quickly accepted, timing his trip to coincide with the Natchitoches Art Colony's spring session at Melrose.[35] His article "Easter Sunday at Aunt Cammie's," published in the *Times-Picayune* on April 22, 1923, gives the details. As the journalist and self-styled southern gentleman met the plantation mistress on the lawn that spring morning, Cammie's manner was "cordial," and her handshake was "firm." Lyle declared that "Aunt Cammie is one of the most charming persons you have ever seen." He happily left the art colony painters setting up their easels before the gardens and followed her as she led him on a tour of her home. He noted various antique pieces that he admired. Of the house itself, Lyle wrote, "Every room opens directly into the open air and the house is full of sunlight. In dim recesses, old mirrors gleam—and the old mahogany shines with its reflected light."[36]

Cammie took him to her favorite room, the library, which at that time was in the garçonnière on the east side of the house. "Here she is in her glory," Lyle wrote. Cammie's library was impressive even at this early stage in its evolution. Dr. Milton Dunn had added his extensive collection to Cammie's growing one, and the result was a large compilation of volumes pertaining to Louisiana and history, including an assemblage of rare and important books. Hundreds of volumes climbed from floor to ceiling in shelves all around the room. "I've always got my nose in a book," Cammie said to Lyle.

Cammie's library also contained her growing collection of scrapbooks. She later explained to journalist and friend Mary Belle McKellar: "As history always fascinated me, such clippings began to grow. When I had finished one book I had innumerable clippings left over and I began to file them according to state sections, or events. As my interests widened, my scrapbooks took new form."[37] In the beginning, she would order her blank scrapbooks from Weis Manufacturing Company, but eventually she would simply produce them herself at Melrose in her bindery.[38] The books vary in size and color, some retaining their original pressboard covers, while others, especially those books she made for her children, might be covered in homespun fabric with their names embroidered on the cover. The number of scrapbooks grew through the years to over two hundred, and Cammie was never unaware of their significance. In 1923 she wrote alongside photos of sentimental homes in Thibodaux: "These books, now 9 vols, represent a lifetime's pleasure of collecting. Some must gather, some scatter—so runs our lives away."[39] Early in their friendship, she told François Mignon that taking up scrapbooking was "lucky" for her: "In my trying days I had to have a 'hobby'—an inexpensive hobby and one that could last a lifetime. Now after half a century of making—and the hobby has fulfilled its purpose."[40]

The books contain personal notations and handwritten oral histories to her children regarding the significance of this home or that one, the genealogy of a person in a photograph, or anecdotal information about what was going on in a picture, as well as notations by authors beside their own articles pasted within. If the scrapbooks started as a childhood "hobby" for a little girl living in rural isolation on Bayou Lafourche, they later became another vehicle for Cammie to preserve and capture what she saw as a vanishing past. Whatever Cammie initially intended when she started her hobby, the scrapbooks have been used by countless writers and students as research tools. Through the years, scholars came to Melrose both to look at the scrapbooks and to consult the rare and historic books and manuscripts collected in her library.

From rural Natchitoches Parish, Cammie spent many hours reaching out to the wider world for information and working to locate books either by Louisiana authors or historically important to Louisiana or the South. She was a member of several historical societies and received their journals, through which she would absorb informative articles and study book reviews and lists that led her then to write to publishers all over the country in search of books that interested her. Cammie's reputation as a collector and historian grew, and soon journalists such as McKellar, Vera Morel of the *New York Sun,* and Saxon were penning columns about the literary renaissance at Melrose. Each article published about Melrose seemed to bring it greater renown and more visitors.

It is ironic that the hobby that created such a diversion for a young, isolated Cammie eventually became the vehicle that brought the world to her at Melrose Plantation, destroying the quiet that she longed for as she got older. Richard Megraw draws a similar conclusion in examining Lyle Saxon's work: through his writing and his later work with the New Deal programs, Lyle was determined to save and preserve old New Orleans and the genteel ways of the Old South, and yet "wrenching things out of local context" may have been a large factor in the modernization that he was powerless to stop. Similarly, Cammie Henry's scrapbooks and her personal library brought vast numbers of tourists, journalists, and scholars to Melrose, closing the gap between the modern world and the Old South.[41]

After touring the big house that Easter Sunday, Lyle and Cammie dashed past the artists painting in the garden, on the side of the house, to the old barn. The building fascinated Lyle, and he noted the huge, hand-hewn cypress timbers with the marks "of the instrument which cut them in shape more than a century ago." Next Cammie showed Lyle the cabin, the building today referred to as "Yucca," where a former slave, Israel Suddath, lived. "Uncle Israel" outlived his wife, Jane, by three years; she died on July 12, 1921. Cammie noted the date in Scrapbook 68 alongside a photo of "Aunt Jane" at Melrose. She also wrote that Jane was "a slave brought from [North Carolina] when a grown girl—sold in [New Orleans] off the old slave block at St. Louis Hotel—bought by Mr. Hypolite Hertzog—and owned by him until freed."[42] Jane and Israel had lived in the cabin at Melrose for the duration of their lives, and author Diane M. Moore notes that as Israel's health and vitality declined after his wife's death, he suffered from palsy and was unable to feed and care for himself. Cammie had her servants help him with his breakfast and dinner, and she personally helped him with his supper in the evenings. When Israel died, Cammie prepared his body for burial, a "final opportunity to render an old servant a service."[43]

When he saw Yucca, Lyle is reported to have declared, "I could write a book right there."[44] Whether it was the cabin itself that inspired him or Uncle Israel is unclear, but it was a prescient statement. Once back in New Orleans, he wrote to Cammie, thanking her for the visit and saying, "If there is any possible way in this world to arrange it, I'm coming back this summer to live in one of your . . . cabins and write a novel—I'm tremendously keen to do it, and the longer I think about it, the more I want to go."[45] He would later call the cabin his own, and though it would be fourteen years later, after four nonfiction works, he did eventually write much of his only novel in Yucca.

With Lyle's visit and subsequent article, word about Melrose spread, and people began to come. Historian Henry Chambers wrote to Cammie, observing: "What a wonderful library that must be of yours! I surely must be given the opportunity to inspect it before long."[46] Chambers had published several historical works, including *The Constitutional History of Hawaii* and *Mississippi Valley Beginnings;* he would publish his best-known work, *History of Louisiana,* in 1925. Chambers had worked as a training teacher at the Normal School in Natchitoches from 1900 through 1902. Through the years, Cammie preserved their correspondence in her scrapbooks; the letters are interesting in that they would have appealed to Cammie's smart sense of humor, with their wry and teasing tone that is much like Lyle Saxon's. Whether it was always Cammie's intention to have a writers' colony at Melrose or whether she got the idea from her involvement with the Natchitoches Art Colony, we may never know, but Chambers was already onto the idea in 1923, when he noted the success of the art colony and suggested, "Why not you establish a Colony for Authors?" He compared writers to prideful lions and suggested there could be "little kiosks scattered about your premises . . . each dedicated to its particular lion," which would not be disturbed during "working hours." He jested that a "feeding trough for the beasts" be established "at some central point on the place, and nearby a spot of lawn where they may paw and prance and sharpen their claws on the turf and talk 'me' and 'what I did this morning' and 'my great work in its present stage' and 'what I am going to do.'" If "kiosks" were cabins and Cammie's dining table could be seen as "a feeding trough for the beasts," then Chambers was quite prophetic in what came to pass at Melrose.

Through the years, Cammie and Chambers continued to correspond; she sent him some plants from Melrose for his garden, and he sent her copies of his books. He addressed her as "Miss Cammie" until the end of 1924, when she became "Hypatia" and he "Herodotus II." Hypatia was an ancient Greek mathematician and philosopher who taught Neoplatonist philosophy in Egypt, while Herodotus was the ancient Greek historian. In January 1924 Chambers sent a copy of a publisher's prospectus of his upcoming volume of *Louisiana History*, which, he noted, included a steel engraving of himself. He quipped, "I shall not mind it if you remove the engraving sometime and tack it up on the wall of your library."[47]

Their letters were filled with discussions of books, plants, and history, but Chambers was also anticipating the future of Cammie's library. In McKellar's 1924 *Times-Picayune* article about Cammie's scrapbooks, she noted that Chambers's letters "continually urge the careful preservation of Mrs. Henry's invaluable collection."[48] He knew the value of what she had collected, telling her: "I think you ought to be considering what is to be the fate of your whole collection, scrap books as well as library. Accident and the Inevitable are always to be looked for and we should never have them come unaware. It would be a pity for that fine collection to get into unsympathetic hands, or be scattered among the unappreciative." He suggested she build a fireproof but picturesque library at Melrose "and install everything you have gathered together, a small MEMORIAL LIBRARY to honor your industry and efforts to preserve the history of our State . . . The idea I have in mind is a sort of shrine where students of Louisiana history may make their pilgrimage."[49]

In one of the last letters Cammie received from Chambers before he died, in 1929, he was still encouraging her to take care of her collection in some permanent way. Directing Cammie toward what she should accomplish before year's end, he wrote: "You are going to set aside some one room or brick fire-proof cabin. You are going to fill all four walls with shelves up to the ceiling. You are going to gather all your books in this one place, group them on your shelves according to subject. You are then going to have someone go over them, book by book, and make a complete card-catalogue, not only by subject but by authors, including individual articles in your bound historical magazines and quarterlies. You will then have a table four feet wide by ten feet long, placed in the middle of the room with chairs around it for the convenience of those who come to research in your peerless collection." He even went so far as to suggest that she hang portraits of Louisiana historians in the room and that she have a catalog printed and sent to editors of historical magazines all over the country, to librarians, and even to the Library of Congress, "so that students may know where they can find certain materials for their study in Louisiana history." He suggested she name her library the "Cammie G. Henry Historical Collection" or the "Melrose Library of Louisiana History."[50] He was close: it would be called the Cammie Garrett Henry Research Center, established at Northwestern State University in Natchitoches, Louisiana.

The final member to join Cammie's inner circle was Ada Jack Carver. Ada, who was one year older than Lyle, attended Judson College in Marion, Alabama, for a year and then returned to Natchitoches to finish her education at the Louisiana State Normal School, graduating in 1911.[51] Afterward she lived in Natchitoches with her parents and worked on her writing. Her first published short story, "Stranger within the Gates," had appeared in the *Winnfield Guardian* in 1907, when Ada was only seventeen years old, followed by "The Ring" in the *New Orleans Item* in 1908.[52] In 1915 "A Pink Inheritance" appeared in the *Designer* magazine, bringing her more recognition, followed by two more published works in 1916.[53] Ada wrote local color stories, many of them set in the Cane River community. While they received glowing critical acclaim, they did not rise to the fame or literary level of those of contemporaries F. Scott Fitzgerald or Ernest Hemingway. Her best work would be done during the 1920s, when she had matured as a writer and had the benefit of the input and feedback of her associates in the Melrose community.

Ada's mother and father both came from distinguished families of jurists, legislators, and attorneys. Ada was a petite, slim woman with stylish bobbed hair, and she wore pretty, fashionable dresses. She met John Snell, six years her senior, while he was on the summer faculty at the Louisiana State Normal School.[54] They married in 1917 in the Shreveport home of her uncle, Judge George Whitfield Jack. At the time, Snell was an army lieutenant and on his way to active duty in Europe, so Ada went with him as far as New Jersey, where she stayed to participate in a writing course at Columbia University.[55]

When John was discharged from military service, he joined his wife in Natchitoches for the birth of their son, John Hampton, in February 1919. They then moved seventy miles northeast of Natchitoches to the small town of Minden, Louisiana, for his business, the Minden Cotton Oil and Ice Company.[56] Ada missed Natchitoches terribly, but she kept busy with her writing, social clubs, and in the evenings, taking long walks with her husband and their young son.

In 1921 Ada was pregnant with their second child when, like Cammie Henry three years earlier, she and John suffered a tragic loss. It is difficult to confirm exactly what happened; there have been slightly contradictory stories through the years, but the consensus is that during one of their customary evening walks, in March on the Saturday before Easter, the family got caught in a sudden spring shower. By the time they returned home, they were drenched, muddy, and cold from the rain. This is where the stories begin to diverge; whether it was Ada or a young servant, someone drew a bath to take the chill off the two-year-old boy. In a brief, unsupervised moment, he ended up in the scalding water and was critically burned. The cause of death was listed as "Burns. Fell in tub hot water. Accident."[57]

The shock of her son's death initiated early labor, and Ada was taken to the hospital. John immediately called Ada's cousin and close friend in Shreveport, Lillian Hall Trichel. It was raining heavily as Trichel made her way to Minden, where she stayed by Ada's side through the birth of her new son and until Ada's discharge. It has been reported that Ada "never again visited alone with Mrs. Trichel," probably because she never again wanted to discuss the loss of her child.[58] She and John buried John Hampton in Minden City Cemetery, where his grave is marked with a stone that says, "Our Little Son John Hampton Snell."

There is no indication or surviving evidence that Ada ever talked with anyone about the death of her son after that, not even with Cammie, who had also lost a son. Struggling with grief and depression, she concentrated on raising her new baby, whom they named David. She and John made plans to build a new home in Minden, a project perhaps intended by John to provide a necessary diversion for Ada's troubled heart. She stayed busy with details for the new house, her writing, and her social obligations in Minden.

As a lifelong Natchitoches resident, Ada certainly would have known about Cammie Henry and Melrose; indeed, her mother, also named Ada, was a friend of Cammie. When Cammie's son Erwin died in 1918, Ada's mother immediately wrote Cammie a lovely and affectionate note of condolence.[59] Although Ada may very well have visited the plantation previously with her mother, it appears her first time there as a writer was the summer of 1924, when she went with Shreveport writer Mary Belle McKellar, a visit possibly engineered by Ada's mother in the hope that it would alleviate her daughter's lingering depression. In advance of the trip, Ada wrote a very formal note to Cammie, who, at age fifty-three, was twenty years her senior: "Dear Miss Cammie—If all goes well we expect to drive down to Natchitoches some time Sunday, and I will phone you then and will make arrangements for my trip to Melrose Monday or Tuesday—as suits you best. It is lovely of you to want me and I can assure you I'm looking forward to it with . . . pleasure. The prospect of Lyle Saxon and his portrait painter is delightful! And you must know what all your friends think of you and your charming hospitality." She signed the note, "Affectionately yours, Ada J. C. Snell, Minden, La.," much more formally than any later correspondence between them.[60]

Once in Natchitoches, Ada phoned Cammie and made plans to come down to Melrose the following weekend. Other guests included Carrie, Lyle, and Edith Fairfax Davenport, a Missouri-born artist who was a cousin of James McNeill Whistler. During that weekend in August 1924, they all drank dark, strong coffee and talked about their various projects. Carrie shared ideas she had for stories about the "Sandhill people" of Kisatchie, and Edith worked on her painting. Ada's past literary success certainly would have been a topic of conversation, though she had not published anything since 1916, a long span of eight years.

That literary dry spell could be partially attributed to her 1917 marriage and her new family as well as her profound grief over the loss of her first child, though she had continued to write and to work creatively, including some poetry, since John Hampton's death. Shortly after Ada's visit to Melrose, her mother sent Cammie a grateful thank-you note saying, "You are a sweet dear friend and have given my girl such a delightful time." She enclosed a copy of one of Ada's poems with the admonition that Cammie say nothing about it to Ada.

To My Little Son
By Ada Jack Carver Snell

So strange it seems to me that he is gone,
I loved him so. Here is his Kiddie Koop,
And here his cart. And over by the door
His Teddy Bear. I hope at first up There
For plaything they'll not give him just a star—
He'd so much rather ride his Kiddie Kar
Across the courts of Heav'n. He sucked his thumb
With such a blissful, sleepy air,
That it will make Saint Peter laugh up There.
And sometimes—just as I loved so to do,
A-singing at my work when life was sweet—
I'm sure from flinging worlds about the Blue,
That God must pause to kiss his rosy feet.[61]

At the end of her stay, Ada's husband and little David came to pick her up. John Snell was charmed by Cammie; Mrs. Carver wrote that he "has fallen into the family way and enjoys your company too."[62] On September 4 Ada sent a letter to Cammie—who was now "Dearest Aunt Cammie"—expressing gratitude for the weekend: "Glad we met . . . You have been constantly in my thoughts since I left Melrose, and I must tell you again that my visit to the Land of the Lotus has meant more to me than I can ever make you know. The thought that I might not have gone terrifies me, really."[63] She told Cammie that *Harper's* magazine had sent a check for one of her stories, and she noted how pretty her "little white house" looked when she got home, even though John had apparently not been the most immaculate housekeeper in her absence, "getting his own breakfasts (as long as the dishes lasted)."[64]

Ada Jack Carver often referred to Melrose as the "Land of the Lotus," and other visitors would adopt the expression in later years. In Greek mythology, the Land of the Lotus was a mythical paradise that one could never leave after eating of the lotus plant. To Ada, Melrose was a place of perfection that gave her great creative inspiration and affirmation. One of Ada's biographers, Mary Dell Fletcher, states that not just Melrose but the Cane River region in general, with its seductive "Lethean quality," were important to Ada's work.[65] In May 1925 Ada wrote to Cammie expressing an almost physical need to visit Melrose, explaining, "It's absolutely necessary to my soul's welfare that I go there occasionally!"[66]

Ada also wrote to Carrie on September 8: "I've thought of you often since our house party—and I'm so glad I know you!" She reported that she had received proof sheets from *Harper's* for her story "Redbone" and a check for three hundred dollars.[67] This would end the dry spell for Ada Jack Carver.

Carrie, Lyle, and Ada would all continue to gravitate around Cammie Henry, who somehow knew how to make each of them produce. They were like-minded people who enjoyed one another's company, and the opportunity to share ideas and support individual endeavors was invaluable. Hungarian psychologist Mihaly Csikszentmihalyi has studied the importance of artists being surrounded by other creative people in their field. He observed that when creative minds come together, they are reassured that "their shared cause is noble," and "the desire to impress the group encourages innovation." Moreover, according to Csikszentmihalyi, the creative personality can immerse itself completely in work to the point that all else falls away. Plush and luxurious surroundings are often more of a distraction than primitive cabins.[68]

It is possible that the same dynamic was in play at Melrose. The cabins were indeed rather primitive, and there were always plenty of creative people on hand. Conditions were ripe for each member of the inner circle to achieve his or her potential. Cammie, Melrose, and the Cane River culture comprised the common catalyst for the successes that lay before them.

3

The Muse of Melrose, 1925–1929

There was activity in every corner of the tremendous garden. No rhythmic
cotton-chopping, no leisurely plowing here! It was fascinating
to watch from the comfortable wide gallery.

—CAROLINE DORMON, "Gardening at Melrose"

After the August 1924 house party that brought Carrie, Lyle, Ada, and Cammie together, a flurry of correspondence began to fly between them. Not only did they all write to each other, but their letters were forwarded to each other. If Lyle wrote a particularly amusing or scintillating, gossipy letter to Cammie, she might forward it to Carrie with "Return!" scrawled across the envelope. Cammie Henry was a woman who preserved almost everything she ever touched, and her scrapbooks are filled with circulated clippings, notes, cards, lists, letters, sketches, and ephemera, such as leaves from a tree at Mount Vernon and confetti from the streets of Paris at the end of World War II.

This was an age when people wrote letters, which are revealing in their composition style as well as their content. Cammie's letters were often a source of much mirth among the recipients, as they were typically telegrammic, with brief, imperative jabs—"Come paint!" One letter in particular from Cammie to Lyle was the epitome of this form: "Iris opening—lovely, lovely—Carrie painting hard—Can't you come see the iris?—Love to have you any day—you can stay all night—of course all varieties are not opening but I want you to see what is open—try to come." On the back of this letter, Lyle wrote: "These are the letters that simply slay me with laughter. I do not believe anyone can imitate her style! I am sending as I know you'd understand I really love them. Please destroy after reading."[1]

The affection between the four friends is obvious in their letters. Cammie and Carrie corresponded frequently, almost always beginning, "Dear Mate," and signed, "Your Mate," regardless of which one of them was writing. Ada's letters were often poetic, sometimes playful, and with the exception of a rare typed letter, written in a nearly calligraphic hand with artful arches, loops, and long tails at the ends of her words. Lyle's were fun spirited, many a time amusing in tone.

Ada usually wrote on heavy stationery with matching envelopes, while Cammie was as likely as not to write on a scrap of whatever happened to be nearby. Both Lyle and Cammie had engraved stationery with a drawing of the big house sketched by Irma Sompayrac printed at the top. Lyle's had his name and the "Melrose, La." designation. Carrie did not seem to use formal stationery; her letters were on plain sheets of paper or scraps she found lying about. They were almost always handwritten, frequently front, back, and sideways on folded paper. In later years she might use formal letterhead associated with her position with the forestry department, engraved with her name at the top, but usually she opted for the casual, informal writing material reflective of her personal style.

By 1924 Lyle was frequently visiting Melrose on weekends as well as spending time in Baton Rouge with his two aunts, Maude and Elizabeth Chambers, whom he supported financially. In 1925 he invited Cammie, Carrie, and Ada to come to New Orleans as his guests for Mardi Gras. The trip would include not only the carnival festivities but also a chance to see some old plantation homes along the Mississippi River. Cammie and Carrie climbed into Carrie's Ford with Cammie Jr. in tow, arriving in New Orleans in the afternoon of Lundi Gras. They settled in at Lyle's, and that evening he took them to dinner and then down to Canal Street to see the Krewe of Proteus parade. The next morning Ada took the train down from Minden and joined them in watching the parades.[2]

On Ash Wednesday morning, Lyle turned his guests over to friends, who took them sightseeing while Lyle worked. The day included a trip to see Belle Chasse Plantation, just south of New Orleans on the Mississippi River, and that afternoon Lyle's friend Olive Lyons had a tea for Cammie. Olive was a dark-haired beauty; she smoked, drank "heavily," wrote poetry, maintained a lovely home and garden, and dashed around New Orleans with the literary New Orleans in-crowd of the 1920s, which included Lyle, Sherwood Anderson, and William Faulkner. Olive was on the advisory council for the New Orleans literary magazine the

Double Dealer, which began publication in 1921, and Lyle served on its staff from November 1921 to May 1922.[3] Olive's husband, Clifford, was the son of a wealthy cotton broker whose family owned a pharmaceutical business.[4]

While it has been rumored that Lyle and Olive may have been more than friends, most consider a romantic relationship unlikely. James W. Thomas, in his biography of Lyle, says that "an untenable situation" had "evolved between the writer and a married woman" and was one of the reasons Lyle often left New Orleans. Chance Harvey, another Saxon biographer, writes that "references to Olive in Saxon's 1925 diary, although ambiguous, suggest a romantic involvement."[5] However, no real evidence of a romantic relationship exists in their letters. She was ten years older than he, and Lyle was close with both Olive and Clifford, noting in his 1925 diary that they brought in the New Year together by trying various and sundry recipes in the bartender guide.[6]

Following Olive's afternoon tea, Carrie, Lyle, Cammie, and Ada went to the movies; afterward they discovered Carrie's car had been stolen. Lyle recorded the event in his 1925 diary: "Police! Lost Car! In the afternoon Ada, Aunt Cammie and I bought Carrie Dormon another Ford to replace hers."[7] Carrie simply had to have an automobile: her livelihood consisted in part of traveling around the state giving speeches or presentations. They contacted Menefee Motor Company in New Orleans, purchased a used touring auto for $265, and had it delivered to Lyle's.[8] Having had enough of New Orleans, the party set out for Melrose the next day with plans to tour several plantations along the way.

Their return trip was a leisurely one. Lyle directed them along the River Road toward Baton Rouge from New Orleans. They stopped for lunch on the levee, Cammie "jubilant over the old houses," and did not reach Baton Rouge until that evening. The group stayed the night and in the morning made a quick trip to tour the Shades Plantation.[9]

But as if the theft of Carrie's car was not enough, about twenty miles from Melrose a wild hog suddenly darted across the road. There was no avoiding the animal. Carrie swerved, and the car careened off the road and down the bank of the river into a low, swampy area, pinning little Cammie partially under the car. They were able to extricate the child, secure another car, and rush to the hospital in Alexandria, where she was hospitalized for a couple of days with a concussion.[10] In his dissertation on Ada Jack Carver, Oliver Ford says that young Cammie was screaming frantically while pinned under the car because the battery had exploded and was dripping acid on her, which resulted in burns that took months to heal.[11] The newspaper account says she was unconscious when finally pulled from under the car.[12] Either way, Ford gives credit to Ada for keeping a level head and organizing the group into lifting the car off of the child. While Cammie stayed at the hospital with her daughter, Lyle took a room at the Bentley Hotel, and John Snell drove down to Alexandria to pick up Ada. Four days after the accident, Joe Henry came from Melrose in the family Studebaker and drove the two Cammies, Carrie, and Lyle back to Melrose.

"How good the cabin looks," Lyle said upon his return. The entire debacle left Caroline Dormon "badly shocked," and Cammie had to reassure Carrie that the accident was not her fault. "You never did a thing in your whole life I would blame you for," she said. "You were magnificent in the tragedy."[13] Little Cammie rebounded quickly, as children often do, and within a week after returning to Melrose, Lyle recorded in his diary that she was "well again and all over the place—how soon we forget terror!" Lyle rested for a few days in the cabin before returning to New Orleans on March 11, and Carrie went home to Briarwood. Ada hoped to use their New Orleans trip as fodder for a new story, telling Carrie: "I am planning a brand-new story inspired by our trip to New Orleans. In fact, it's a tale of the Old French quarter. I'm in fine writing trim."[14] If Ada ever finished this story, it has not been found. The return to her little house in Minden made Ada acutely feel the separation from her Melrose circle, and she wrote to both Cammie and Carrie imploring them to bring Lyle, "when the car is in shape again," to Minden: "My house is sweet and shining, and ready for you when you can come and see me. Really, you must do it."

Just over a week later, Lyle was back at Melrose with Olive and Clifford Lyons. Spring at Melrose was spectacular as the early bulbs pushed through the earth, and Cammie's beloved irises unfolded their blooms in a riot of purples, pinks, whites, and yellows. Cammie was such an iris enthusiast that their beauty sometimes eclipsed her vocabulary; she once wrote to Carrie in an attempt to describe a particularly beautiful yellow iris but found herself without the right words, so she sewed a tiny scrap of yellow fabric to the paper with three stiches and wrote, "Exact shade!" next to it.[15] The pink magnolia trees and the wild dogwoods were in bloom, and spring rains nurtured the young vegetable garden; the plantation was glorious. Melrose was always at its best in the spring, with colorful blooms at every turn, and warm, mild temperatures coaxed everyone out of doors to take advantage of the longer days. The slow, easy pace of the plantation allowed for extended afternoon drives along the Cane River, leisurely walks in the garden, and cold drinks on the gallery at Lyle's cabin. On one afternoon, Olive and Lyle drove to Briarwood to visit Carrie.[16] At night, after fetching ice from the big house, Lyle, Olive, and Clifford would build a fire in Lyle's cabin to chase away the evening chill and spend hours talking, gossiping, and laughing until the early morning. Cammie seldom took part in these late-night sessions, preferring to retire to her room before her own fire to read, write letters, and work on her scrapbooks. Shortly after the Lyonses' visit, both Cammie and her mother found themselves in bed for a week with the flu. Lyle took advantage of the quiet days to read, write letters, and work in his cabin.

It was summer before Ada made her way back to Melrose, although she did occasionally get to Briarwood to see Carrie. Through the spring, Ada reveled in her own little garden and house as she worked on her stories and wrote letters. Her yaupon hedge, which she had transplanted from Melrose, was thriving, and in a lighthearted note to Cammie, she lamented the fact that there was a mockingbird outside her window in the hedge that "wakes me up every morning between four and five—and I try to be thrilled—for his song is lovely. But, I am not Carrie . . . and I usually feel like throwing something at him."[17]

Ada's trips to Briarwood almost always included her son David, who as an adult wrote about those visits for *Smithsonian* magazine in his tribute to Dormon after her death. He described driving to Briarwood with his parents, almost always with a block of ice tied to the front bumper of their car, a luxury to Carrie and Virginia. Ada and Carrie would chatter about friends, while young David would look for doodlebugs or go exploring. Carrie loved teaching everyone who came to Briarwood about the various plants and flowers that she cultivated there, but she especially loved teaching children, and David learned much about nature at her hands.

As summer 1925 approached, Lyle announced his plans to stay at Melrose the entire season and write; in the fall he would have to make a trip to meet with his publisher in New York, he explained, but would return to the cabin after that. "Nothing could please me more than this," Cammie told Carrie. "I want him to be happy and write—the two are synonymous. I want his greatness to come to him in the cabin—his cabin."[18]

Biographies of Lyle tend to depict him as a charming bon vivant on the exterior but internally tortured and depressed. It appears that during those early years in his friendship with Cammie, he was, for the most part, quite happy at Melrose. He seemed to live in two worlds. At Melrose he could relax, escape, write, and visit; he was a celebrity there. In turn, he would occasionally have to return to civilization—to Baton Rouge, New Orleans, or New York; his bread-and-butter writing depended upon it, and his two aunts depended on him. John Bradbury contends that for "young, artistically-minded Southerners, the North offered the same lure that Europe suggested to Fitzgerald" and the other expatriates.[19] Perhaps this was true for Lyle; he needed the stimulation of city life and his literary friends there but still yearned for Melrose. At times his letters and diaries express boredom with Melrose and its isolation, but they are especially few in the early years. At that point he still had his dreams of a novel before him and his common interests with Cammie in old homes and plantations, and the old ways of life in the South still infused his writing. He was taken with the novelty of Cane River and its people. His cabin at Melrose was the embodiment of that fascination and a sort of muse for his fiction.

Lyle's cabin, Yucca House, is one of the two oldest buildings on the Melrose property, along with the African House, or the "Mushroom House," as Cammie called it. Another Melrose structure renamed by François Mignon, the African House is a small, two-story, two-room hut of unique construction that resembles a giant mushroom. It has a large, overhanging roof that conceals the upper story. The entire edifice is built without any nails. Mignon contrived the fantastic story that it was built by Marie Thérèse Coincoin in her longing and homesickness for the African continent. He wrote that she "felled the giant cypress trees" and seasoned them "presumably for a six year period while the bricks were being baked for the first story of the structure."[20] Mignon dated the structure to 1750, but current research belies all of that. The African House was likely built in the early 1800s, and researchers contend that there is nothing African about it. More likely it is of the style of French farmhouses.[21]

Yucca House was constructed of *bousillage* daubed between upright logs, and current research dates it to the 1820s.[22] There is a large gallery, or veranda, both front and back, which at the time of construction ran all the way across the width. Inside there are four rooms, two of which Lyle used as a bedroom and a sitting room. The rooms were at some point painted white, and in the sitting room there were bookshelves around the fireplace that went from floor to ceiling, which Lyle filled with his large collection of books and objects that he found interesting. There was a handmade rug in front of the fireplace, where he had custom andirons that bore his initials, *L* and *S*. His typewriter was in the sitting room and faced a blank wall, his back to any tourists or visitors that might pass by in their strolls through the gardens.[23] The doors on the front and back open onto the galleries and face north and south. Lyle planted banana plants, which grew quite large; he mentioned them occasionally in his letters when he was homesick for Melrose and the cabin.

In June, Ada and David visited Melrose; Carrie met them there, and after a visit with Cammie and Lyle, the three returned to Briarwood to spend a few days. It was not uncommon for Carrie, Ada, and Lyle to share and brainstorm stories when they gathered at Melrose, then they would read them aloud and discuss them with Cammie. Sometimes one of them would suggest a random topic on which all three would then write something.[24] One such example is the centaur. It is not clear how they decided on that subject, but there is at least one earlier reference to the mythological creature—in Ada's story "Redbone," which had been published in February of that year: Baptiste Grabbo is compared to a centaur as he rides into town to celebrate the birth of his son. The beast then appears in a June letter to Cammie signed jointly by Ada, Carrie, and David. The letter has the hallmarks of Ada's airy composition style, which indicates she was likely the typist: "The hoodoo is at last dead and buried. We came all the way back without mishap, but alas! The witch's moon was gone, so there have been no more Isadora Duncan performances by its silvery light . . . Last night it came a delicious little shower. We had gone out to walk the 'three miles' to the front gate so the shower caught us. We confidently expected to see satyrs, fauns, dryads—perhaps even centaurs—come springing out from woodland coverts. The pines were wet and sweet and the eaves dropped deliciously all night long."[25] The letter reflects the silly kind of fun that often developed when Ada and Carrie got together, and one can only imagine the amusement it brought to Cammie and Lyle back at Melrose.

Shortly afterward, Ada wrote again and urged Cammie to have Lyle keep working on his centaur: "Tell Lyle to be sure to work on the centaur story. I cannot forget it and his irresistible treatment of it. Tell him . . . we must also do a story some day of a centaur who (or which) comes out of the woods and kidnaps a woman and takes her to his haunts where she reigns as a queen. I was reminded of this at Carrie's when that last night she and I got caught in the woods in the rain and heard hoof-beats!"[26]

They were still talking about centaurs in the fall, when Cammie asked Ada if she was familiar with the "Centaur Bookshop" in Philadelphia, and Ada affirmed, saying, "The ad with its prancing centaur always fascinates me."[27] The centaur motif was central to Lyle's story "The Centaur Plays Croquet," which was written sometime before the end of 1925 and included in the 1927 volume *The American Caravan*.[28] Lyle was working on his story as early as June, when he wrote to Carrie that "the centaur continues to gallop . . . wildly. I worked a while on him again this morning."[29] By August, Carrie had produced her centaur, which she sent to Lyle. "I'm sending you your neatly typed Centaur-monologue," he said. "I've just read it again and I think it has some lovely cadences in it . . . I wonder if you and Ada and I will ever put our Centaurs together in some distant future? I like to think so."[30]

"The Centaur Plays Croquet" is classic Lyle Saxon, and the Melrose influence is evident. The story, told in a series of "affidavits" and court testimony, centers on Mrs. John David Calander of Mimosa Plantation in Louisiana, drawn from the names of Ada's husband, son, and a play on her maiden name. Ada Calander "preferred spending her time in the woods and fields," much like her real-life counterpart. Ada Calander is, of course, not literally Ada Carver, but the inspiration is clear. As the story goes, Ada Calander and her husband lived on her inherited plantation, a "lonely place, seeing few people and entertaining visitors but rarely." (Melrose was an inherited plantation but was certainly never at a lack for visitors.) Ada, who was reared in a Catholic orphanage, has rejected Christian religion and adopted a fondness for mythology, publishing poems in *Harper's* magazine, all quite pagan in nature and somewhat scandalous, but her good Baptist husband stands by her side despite the "busy tongues" on Mimosa Plantation. Before long, Ada brings home a pet centaur, whom she calls "Horace"; she writes a "prose poem" called "The Centaur Plays Croquet," which "was found among her papers after her death."

James W. Thomas, in his biography of Lyle, analyzes the "sexual implications" of the poem and draws comparisons between the phallic mallet and the wicket of croquet to the pet centaur and Ada Calander. Their relationship is abnormally close, much to the dismay of the husband. Ada Calander and Horace spend their days going off into the woods, and "together they would return at twilight." The centaur takes meals with the family, and Ada dresses him in her husband's clothes. Eventually, the whole situation becomes unbearable, and Mr. Calander mentions one evening in Horace's presence that he has seen a beautiful white mare outside. Horace sheds his clothes and goes out the window, and Ada begins to lose her sanity, until finally Horace and his white mare are brought back to the house. It is not long before the white mare dies a horrible, painful death, and Ada Calander is suspected of poisoning her. This is all too much for the citizens of Mimosa, who band together and come to the plantation with the mayor and the Baptist preacher, but Ada and Horace escape before they arrive. Poor Mr. Calander is so distraught at his wife's behavior and abandonment that he commits suicide, but he leaves his estate to his wife nonetheless. She eventually returns, heavily veiled, and stays isolated at the plantation, refusing all callers. Horace dies and is buried under a magnolia tree on the plantation, which is leveled after Ada Calander's death and a country club built in its place.

The story was meant to be fun and whimsical, but in his biography of Carver, Oliver Ford states that Ada "took offense" to the story, "or some such allusion which she considered unseemly," and did not complete her own centaur entry for their game.[31] She never mentioned the story or the exercise in her letters that followed. In fact, when *The American Caravan* was reviewed by the *New York Times,* it was criticized for the lewdness of some of the stories. Reviewers complained that some of them showed "a very strong leaning . . . to the unsavory, not to say bawdy."[32] It is certainly possible that Ada Jack Carver either found the story just a bit too far out for her sensibilities or resented its obvious personal references.

Carrie stayed out of the fray, although she did write her own centaur poem, entitled "The Captive Centaur" and subtitled "In answer to Lyle Saxon's 'The Centaur Plays Croquet.'" The poem is from the perspective of the centaur and concludes with the lines:

> Today I saw wild deer sweep past me on the hillside,
> And disappear in the thickets of red-stemmed poison sumac,
> Their Pan-hoofs cleaving rich black woods-loam,
> As they passed their silken sides brushed mine—you were forgotten.[33]

Later that month, Carrie's article about blueprinting techniques was published in *Holland's Magazine* and brought some much-needed cash to her coffers. Carrie had been using her blueprinting technique for several years, and in October 1924 *American Forestry* had published her article "Tree Studies in Blueprint," in which she advocated blueprinting as the simplest and cheapest way to "get a record of leaf, blossoms, and seed."[34] The process requires only an inexpensive frame and blueprint paper, which can be found at "any house that handles drawing materials," she explained. The specimen is arranged on the glass, the paper spread over it, and then it is exposed to sunlight for a few minutes. The resulting image should be washed in water and left to dry. She usually used a solution of chromate of potash, but when working with young children, water is fine, she explained. Over the years she had accumulated hundreds of these blueprints, finding they preserved the image and delicacy of specimens, losing only the original color. Her 1925 article in *Holland's*, "Blueprinting the Wildflowers," extended the idea as a calendar project, suggesting that one record specimens throughout the year and compile them into a yearlong recollection of the beauty of nature.[35] The prints were quite popular, and in a letter to Lyle, Carrie told him she had sold "nearly three hundred dollars worth of blueprints . . . I really believe a book of them would sell."[36] By June 1926 Carrie had begun to collect some of her favorite images with the thought of publishing them, as she had amassed over one hundred species of wildflowers in blueprint. She told Cammie that "everybody seems to like my prints," and she was hoping to publish them and eke out enough money to quit teaching.[37]

Cammie suffered from ongoing dental problems, and by 1925 they were causing her such pain and headaches that she was driven to distraction. She suffered mightily throughout the early part of the year with various infections that caused pain, headaches, and a general feeling of malaise. In January she wanted to go to town to get her teeth taken care of, but because the dirt roads were impassable, she simply had to deal with it, although she did try to keep good humor about it. In a letter to Carrie, she wrote: "Come on Mate—let's hit the road (teeth or no teeth for me). I just have a longing to roam French N.O. 'Vieux Carré'—we can eat bread & water & feed the soul—Joe is here to stay with Mother—I do believe we could run—If I could first get the offending teeth out."[38]

In May she finally had several teeth extracted and then had to wait for the swelling to go down and healing to occur before she could get plates. When the dentist told J.H. that it would be three months before his mother could get her teeth, Cammie "immediately threw a fit," saying, "I won't be incapacitated for that long—bound to be a way out—I'll see that there is."[39] And in fact, by June she spent "eight terrific days at the dentist" getting her plates so that she could go in July to see Bobbie. Lyle wrote to Carrie that "Aunt Cammie and Sister [Cammie Jr.] and Joe went in the Studebaker to Shreveport in order for Aunt Cammie's trip to the dentist to be an accomplished fact instead of a subject for conversation."[40] Her absence left Melrose very quiet, although artist Alberta Kinsey was there painting.

Alberta was born in Ohio in 1875, came to New Orleans in 1918, fell in love with the city, and made it her home. One of the first bohemians to move into the Vieux Carré when it was still a less than desirable neighborhood, she was perhaps best known for her paintings of quaint courtyards lush with bougainvillea, greenery, ironwork, and unique architecture. She and Lyle were close friends and ran in the same circles, and she was one of the founding members of the Arts and Crafts Club in New Orleans. As he did with many of his friends, Lyle introduced Alberta to Cammie, and she began painting the scenery along the Cane River. Her presence at Melrose gave Lyle some diversion and perhaps an opportunity to procrastinate further in writing his novel, as he took time away from his typewriter to drive Alberta to various sites to paint cabins, magnolias, and churches. She would sit under the trees, smoking a cigarette, and paint for hours; then she would take her paintings and sell them for a nice profit. In a 1926 letter to Carrie, Cammie exclaimed, "Do you know Kinsey sold a thousand dollars' worth of pictures in Shreveport?"[41] Alberta developed a very close, enduring friendship with Cammie, saying of her, "She was the most understanding person I have ever met." If a painting was not going particularly well, Alberta would take it to Cammie. "She did not know a thing about painting," wrote Alberta, "but she understood my problems and would say the right thing to make me want to go back to work."[42]

While in Shreveport with the dentist, Cammie had hoped to take a trip over to Minden, where Carrie was visiting Ada. Whether she refrained from going because of her dental troubles or, as she later said, "I knew you two were working together," there is no way to be sure. By early July, though, Cammie was off to Camp Meade in Maryland to visit Stephen, who was serving as an instructor to military students in the army. She and Stephen then headed to see Bobbie in Tennessee. In her absence from Melrose, Lyle was appointed to stay and keep an eye on the place and on Leudivine. "So glad he is here to be with mother—I feel better—I'll come back saner and better," she wrote Carrie.[43] Cammie never left Leudivine at Melrose without worrying about her, regardless of who was there to look out for her.

Cammie and Stephen took a picnic lunch to Mount Vernon one afternoon, fulfilling her lifelong dream to see George Washington's home. She wrote to Carrie: "Of course I am entranced with the historic land marks—I'd go a million miles to see Lee's, Washington's homes—Mt. Vernon! It far exceeded all expectations!"[44] Cammie took pictures and brought back leaves from a tree at Mount Vernon that she pressed into one of her scrapbooks.

At the end of July, the two drove from Baltimore to Philadelphia and then headed to New York. Cammie wired ahead to Lyle's friend Noel Straus, a journalist and music critic, who met them and took them sightseeing. Lyle reported to Carrie: "Aunt Cammie is still gallivanting around in the Frozen North. She spent a week-end with Straus, you know! . . . In one letter she talks as if she were coming home immediately; in the next she says vaguely, 'in ten days or so, I'll be moving onto Murfreesboro' . . . Mrs. Garrett, vexed, says 'Well! I do wish she'd make up her mind, or stop talking about it!' She's secretly pleased, however, that 'daughter' is having such a good time. I'm perfectly devoted to Mrs. Garrett; like her better all the time."[45]

While the stream of visitors slowed somewhat in Cammie's absence, the plantation was still a busy place. Whether she was there or not, tourists driving the River Road would often stop to admire the gardens, which were glorious year round. August was cotton picking time, and the gin was running steadily. This activity was yet another diversion from his novel for Lyle, who spent a great deal of his time writing long, chatty letters to Carrie and taking "countless pictures of everything." "The plantation is at its most plantation-ish stage," he reported.[46] One subject he particularly liked to photograph was the baptism rite, with its shouting and celebration. In early August plantation employee "Mug" was baptized at the church on Little River, and Lyle was there with his camera to record the event, the photographs of which are preserved in Cammie's scrapbooks. On another occasion he photographed the Melrose employees on laundry day with their galvanized tubs over blazing fires, tables, baskets, and washboards. Cammie's handwritten caption for the photos says, "Wash day at Melrose, 1925—It takes Lyle to get them in characteristic pose."[47] Cammie was always careful to note everyone's name under the pictures, a testament not only to the closeness between her and her friends but also to the affection and regard she held for the people who worked for her.

Lyle recounted one visit Ada made to the plantation that summer in a *New Orleans Times-Picayune* article he wrote about a year later, after her success with her play *The Cajun*. He wrote that they spent a long evening sitting on the gallery of his cabin talking about her achievements and especially the excellent reception of her story "Redbone." Ada was finding it all a bit overwhelming, telling Lyle that success "makes you a little frightened, and a little appalled." "Now I feel dissatisfied. I want to go on and do better things," she explained.

> "Not that 'Redbone' is not my best. It was. It was the very best I could do at that time. But I want to write other things. I want . . ."
> "Well," I said, laughing, "what do you want?"
> "I don't know, exactly," she said. "I want the moon, I suppose!"[48]

They discussed what makes good writing, and Ada professed her reliance on sincerity and restraint. Style is not essential, she said: "Say a thing the way you feel it." What is telling about the *Times-Picayune* article is the obvious affection and respect that the established writer, Lyle Saxon, had for his friend Ada, saying she "is writing as fine short stories as any written in the United States today."[49] While Ada was clearly a gifted writer, anxiety about her work and a somewhat sheltered life may have ultimately stifled her creativity. She never made the pilgrimages north, as Saxon did, never left the country, as the expatriates did, and never really left North Louisiana. If she wanted to go on "and do better things," it is never manifested, and at this period in her writing career, she is at her peak.

Lyle spent the summer of 1925 working on "Cane River" and finished it by the end of August. He had been writing the story off and on through the years, first with the title "Trick Nigger." It is about a wild black woman named Suzy who flaunts her affair with Big Brown to her husband, Babe Johnson. Suzy wears audacious red beads, a gift from her lover, and plays the harmonica on her porch; she is quite scandalous. In a letter to Carrie in early August, Lyle says: "I've discovered a real Susie! How nature does follow art . . . Her name is Nina, and she lives out on Little River . . . she is skinny and scrawny, with huge lips, and a wild face, and she plays the harmonica!"[50] As pleased as he was with his story, he was frustrated with his progress on his novel and noted in his letter that it had been "a whole year that I've been away from the *Picayune* . . . and have accomplished nothing . . . It makes me sick to think about it."

While Lyle struggled with his novel, Carrie returned to teaching that fall, which distressed Cammie to no end. "I can't *bear* to *think* of you teaching," she wrote to her friend. "Lyle says your last story 'Clothes Line' is splendid. So does mother. I hate to think you'll be too tired to write."[51] Lyle had raved over Carrie's short stories, and Cammie wanted desperately for her to have the freedom to write and paint as she knew Carrie could do. But the perpetually broke Carrie had to earn a dollar, and so back to teaching she went. In September, Cammie again wrote, "I hate to see you teach—didn't want it," she said. "Wanted you free to write—drudgery kills inspiration."[52]

Cammie was out of sorts at that writing, missing Lyle—who in his absence from Melrose was "silent as the grave"—and concerned as usual about her mother's health. She also had to send Cammie Jr. back to Natchitoches for school and to board with friends. Cammie Jr. left Melrose after every school vacation "with tears in her eyes," homesick as ever.[53]

Early September brought Ada and son David back to Melrose. It was unbearably hot, and there had been little rain. Cammie and Ada spent the evenings reading and talking; Ada no doubt replenished her creative spirit with the visit, sounding story ideas off Cammie. If the two ever actually spoke of the death of Ada's first child, this visit may be when it happened. They apparently had a deep and heartfelt talk, perhaps about Ada's indignation about the centaur exercise or maybe about the loss of her son. Upon her return to Minden, she wrote to Cammie: "I have missed you so much! Since my return (in Natchitoches I was much too busy to think and in Melrose I never think—it's quite impossible! I just live!) I've done a lot of thinking . . . and I have taken all the heartache and indignation that was stored up in me and have weighed it against your love and marvelous understanding, and am fully repaid. Aunt Cammie, I really wish I were your age (or maybe a little older, for you are awfully young, you know!). And I wish I had behind me, to my credit, all the useful, lovely life that has been yours. Then I would, perforce, have earned my freedom and could 'let the world go hang.' Well, just wait till we are seventy-five, you and I!"[54]

In that letter, Ada mentions a new story she has started called "Three Gables on the Cane," but a story was never published under that name; perhaps it was one of the several manuscripts that Ada began but never completed to her satisfaction. Another story mentioned several times in her letters is "Wedding Cake." In a November letter, she writes Cammie that it had been rejected twice. The story was never published, nor has it ever been found.

Throughout the fall of 1925, Ada sent one letter after another to Cammie expressing how "homesick" she was for Melrose. The two exchanged letters about books they were reading, and Cammie sometimes loaned books to Ada, who had "a little shelf sacred to Melrose" in her Minden home, "and no other books desecrate it."[55] In November, Ada sent a letter stating she had read "five of the books you sent me," and in addition she lists five other novels, as well as "some books on psychology" that went along with a course she was taking at the university. Cammie noted in the margin, before forwarding the letter on to Lyle, "Some reading she's done," a reflection of the deep regard and admiration she held for Ada. In the same letter, Ada described a recent trip to the Louisiana State Fair in Shreveport with her husband and son and how they spotted some nuns taking a chance on the Wheel of Fortune. "One of them won a baby doll" and was very pleased, she wrote. "It was a delicious, incongruous picture and one that Lyle would have enjoyed." At the bottom of the letter, Cammie wrote to Lyle: "Isn't this Ada—wouldn't you have enjoyed seeing the nuns—Ada gets so much out of life!"[56] Lyle's response to Cammie was typically mirthful: "Your letter came the same day the box did, and this morning the other letter with Ada and Carrie inside! What girls they are, and aren't the letters absolutely characteristic? I chortled with glee over both of them. They are so completely themselves. I can picture Carrie and her log fire at Briarwood, and of her happiness in the Autumn woods . . . and I can see Ada following the nuns about, at the Fair. What a lovely thing to have happen! I'm quite envious."[57]

Lyle's fascination with nuns was evident from his early days in New Orleans as a *Times-Picayune* reporter. In February 1922 he wrote an article about the Ursuline order for the paper; he also included a history of the nuns in *Fabulous New Orleans* and later in the Federal Writers' Project Louisiana guidebook. Whenever artist friends would visit, he would take them to see the nuns, who lived for years in a large convent on Dauphine Street but moved uptown to State Street and Claiborne Avenue in 1912, when the levee board decided that the Mississippi had eroded so much of their land that the levee needed to be realigned. Part of Lyle's interest could be attributed to the historic associations of the order; at one time the nuns were housed in the home of Bienville, after they first made their journey from France to New Orleans. Lyle's 1922 *Times-Picayune* article included the first photograph of an Ursuline in a newspaper permitted by the order. The Ursulines were cloistered but not isolated and received visitors often. Almost certainly, Lyle would have entertained Cammie, Ada, and Carrie with stories about the nuns whenever he was at Melrose.

Lyle's presence at Melrose brought many of his famous friends to the plantation through the years. In a letter to Carrie dated January 26, 1926, Cammie reported that Lyle and New Orleans journalist Natalie Scott were at Melrose; "only a flying visit," she wrote, adding that "Bill Spratling came yesterday." It may have been on this 1926 visit that Spratling did five sketches he called "Cane River Portraits," which *Scribner's* magazine published in 1928. His subjects included Madame Roque and three members of the Melrose staff: "Messaline," "Duggy," and Henry Tyler. Spratling made arrangements with Cammie to return at a later date to sketch Melrose as potential material for the book on which he and Scott were collaborating—*Old Plantation Houses in Louisiana,* published in 1927. Spratling and Saxon's friendship dated to 1922, when Spratling came to New Orleans to teach architecture at Tulane. He quickly became part of the art colony in the Vieux Carré and developed friendships with Scott, Alberta Kinsey, Sherwood Anderson, and William Faulkner, among others. Spratling and Faulkner published *Sherwood Anderson and Other Famous Creoles* in 1926, a small book of caricatures of their bohemian friends in the Quarter. Spratling returned to Melrose several more times before he left Louisiana in 1929 for Taxco, Mexico, where he began designing silver jewelry, which in large part revived a silver industry in Taxco. Saxon and Spratling remained friends and exchanged letters throughout the rest of their lives.

Meanwhile, in 1926 Lyle's short story "Cane River," about the wild Suzy, was accepted by the *Dial* for publication that March. He wanted to leave the *Times-Picayune* behind him and work in fiction, so the success of "Cane River" must have been encouraging. While Cammie rejoiced, she also worried that

Lyle's continuing financial troubles would cause him to have to return to the paper. In January 1926 she wrote to Carrie: "Dear precious Lyle—will he have to go back on the paper? Oh—I hope not—if he does he is swallowed up for me—I'll never have him again."[58] She wanted him at Melrose, in the cabin, writing.

In February, Shreveport journalist Mary Belle McKellar came to Melrose to work on an article about Cammie and her scrapbook hobby. Published on Sunday, March 14, in the *Times-Picayune*, it was typical Mary Belle: enthusiastic, effusive, and quaint, as she wrote about "friends who pilgrimage hither" to Melrose. Mary Belle worked often at Melrose and would stay in one of Cammie's writers' cabins. When away, she sent Cammie typed letters, often pages long, single-spaced, and filled with gossipy news of their friends.

At the time of the article, Mary Belle reported that there were twenty-five scrapbooks "bearing on their broad backs hand-lettered titles of the parishes." Mary Belle accurately predicted they would be "a priceless and patriotic service to students of history for generations to come." Many contain firsthand accounts written by the authors of some of the articles as well as marginal notations, corrections, and elaborations. For example, Cammie had preserved many published pieces about Natchez, Mississippi, by Edith Wyatt Moore; several notations and explanations by Moore are jotted alongside these articles, offering further insights.

Mary Belle's article was certain to attract even more curious spectators and students of history to Melrose. Besides covering the scrapbook collection, it described Cammie's home to be almost like a hostel, with beds in every room and hammocks strung up across the galleries for travelers to rest their weary souls. "Melrose draws the great and to-be great as the moon draws the tide," she wrote.

On occasion someone would contact Cammie inquiring how much it would cost to stay at Melrose; anyone could visit the gardens, but an invitation to stay was reserved for friends. Cammie never charged anyone to stay there, only requiring that they work at their craft. Future Melrose guests would include Texas folklorist Dorothy Scarborough, who visited Melrose in 1925 and talked with plantation residents Uncle Israel and Aunt Jane as well as Cane River mulattoes to gather material for her book *On the Trail of Negro Folk-Songs*.[59] In 1930 New York novelist and illustrator Rachel Field visited Lyle at Melrose; that year Field's children's book *Hitty: Her First Hundred Years* won the Newbery Award.[60] Natchitoches novelist and *Times-Picayune* journalist Gwen Bristow visited Melrose in

1933; in 1937 she would publish *Deep Summer,* the first in a trilogy on plantation life.[61] Author Roark Bradford visited Melrose several times in the early 1930s; a friend of Lyle's, he also wrote for the *Times-Picayune,* and he published short stories in various literary magazines. Bradford achieved great success with his collection of folktales titled *Ol' Man Adam an' His Chillun* (1928), which was later adapted for a Broadway play called *Green Pastures.* Some of the luminaries who came to Melrose as friends of Lyle also developed a friendship with Cammie, such as illustrator Edward Howard Suydam. His letters preserved in her scrapbooks are all on heavy, cream-colored stationery with his perfectly vertical calligraphic script.

All of the visitors were party to an unusual custom at Melrose. Mary Belle McKellar may have been one of the first to write publicly of Cammie's legendary tablecloth custom: "Those chatty, convivial meals! One of the fascinating souvenirs of Melrose was begun at a Melrose dinner. There is a tablecloth spread for special occasions when guests of particular note—whether in the eyes of the world or the hearts of the family—here found their way round the curves of old Cane river for this enchanted spot. As the black coffee is brought in, each guest who has not left his signature on this white cloth must do so, and later it will be embroidered in vari-colored threads."[62] Mary Belle called the tablecloth "a sort of hand-embroidered guest book." Cammie herself described this tradition in a 1922 letter to Carrie on the occasion of the birthday of Dr. Ragan, the local country doctor. She had each of her fourteen guests "write their names on the tablecloth—now I'll outline and preserve for another great day."[63] Through the years, there were several of these tablecloths, and the tradition even continued for a while after Cammie's death, carried out by her family.

In her article, Mary Belle identified what would be Cammie's lasting legacy: her unselfish support and encouragement of others to create and produce, whatever their aim in life may have been. Cammie's letters to Carrie most obviously exhibited those traits. Year after year, she implored Carrie to come to Melrose and write, paint, rest. Carrie had ideas for some short stories about Melrose that Cammie urged her to put to paper. "Write the Melrose stories before the impression becomes too faint," she stressed, and then, as if she knew her own pleadings would not be adequate to slow Carrie down long enough to write them, she paraphrased Robert Burns: "Inspiration is very fleeting—short lived—'like snowflakes in a river—a moment white—then gone forever.'"[64]

Carrie had talked about these stories as early as 1925, when she asked Cammie: "Do you suppose I am big enough to ever write a series of stories about it? Could I get all the sweetness and the humor of it? I want to be the one to do it, for I truly don't believe anyone else loves it as much as I do."[65] Carrie wrote many "Sand Hill Tales," and it appears she started to write some of "The Melrose Stories," but she never sold a word of her fiction or poetry. She was deeply involved in her conservation work and public speaking, and biographer Fran Holman Johnson writes that Carrie "literally worked day and night. She designed and implemented programs and activities, always looking ahead to reaching more people and teaching more nature subjects."[66]

In April 1926 Mary Frances Davis of the Louisiana Normal School in Natchitoches brought her drama class to Melrose. Fresh on the heels of her successful one-act play *The Cajun*, Ada was invited to speak to the group. The women lunched in the screened-in summer dining room, which was decorated with flowers from Cammie's garden. Ada delivered her talk, "The Making of a Play," the same speech she had given in Shreveport for the Woman's Department Club. The group then toured the gardens and posed for a photo for the *Shreveport Times*.

Ada's play had won the Shreveport Little Theater playwriting contest in February, defeating eighteen other entries from all over the country. She won a one hundred–dollar cash prize, and the committee unanimously decided to enter her play in the prestigious Belasco Cup competition in New York, a competition intended to bring small-town, amateur plays to a more formal stage. At the end of the month, a very excited Ada and the entire cast of *The Cajun* chartered a bus and traveled to Melrose so the actors could meet Cammie, who helped them select antiques and other items to be used as set decorations for the New York staging of the play.

Cammie gave the cast a tour of the plantation, and then they piled into cars and toured Isle Brevelle, stopping in to visit Cammie's neighbor Madame Aubert Roque, the granddaughter of Augustin Metoyer, eldest son of Marie Thérèse Coincoin and brother of Louis, who established Melrose. Roque regaled the New York group with stories of Grandpère Augustin and his dedication of land for St. Augustine Catholic Church on Isle Brevelle. She showed them his life-sized portrait, which still bore jagged slashes made by the swords of the Yankees during their retreat from the region during the Civil War. A friend and neighbor of Cammie ever since she arrived at Melrose, Roque once made a quilt for her, "each block a different pattern . . . and each pattern a special name."[67] Cammie would often have her driver take Madame Roque to Sunday services at St. Augustine, just across the river. Through visits with Roque, she learned the historic and cultural value of Isle Brevelle and recorded many of Madame's recollections in her scrapbooks.

Ada felt the trip was a great advantage to the presentation of the play, as it gave the cast a sense of the atmosphere and "local color" that she hoped would come through in the production. Among other items, they secured the loan of a turkey fan, andirons, a rocking chair, a handmade basket, a mantle clock, curtains, and Spanish moss. Most of these items traveled to New York for the opening of the play and also appeared in the Natchitoches presentation later that spring.[68] In the New York competition, Ada's play ultimately won two first-place votes but came in second place overall, losing out to the Dallas Little Theater group's play.[69]

The Cajun depicts the plight of first cousins Pierre and Julie, who are on the verge of marriage when the First Cousin Law, prohibiting first cousins to wed, was enacted in Louisiana in 1900. It was a common practice in rural Cajun Louisiana, and Ada's play captured the pathos and tragedy of the situation in a simple and heartrending way. After its New York premier, *The Cajun* returned to Louisiana and was staged first in Shreveport on May 18–21, at the Woman's Department Club, and then in Natchitoches at the end of the month. By then, Ada's reputation was rightfully glowing. Her play presented a little known facet of Louisiana culture to the nation perhaps for the first time.

One of the attendees at the New York production of *The Cajun* was Ada's friend and the Natchitoches Art Colony cofounder Irma Sompayrac, who was working in New York as a freelance artist. Immediately after seeing the performance, Irma dashed off a telegram to Cammie in which she raved about the wonderful reception of the play. It received "the greatest applause of the evening. The properties from Melrose plantation made the setting," she said.[70]

While entertaining Ada in New York that spring, Irma was also planning her June wedding to David M. Willard Jr. of Long Island, a real estate developer she had met the previous year. The couple planned to honeymoon at Melrose for a few days. Very excited, Cammie broke the news to Carrie: "Hold your breath! I'm to have a bride and groom in cabin next week—Irma Sompayrac married . . . they chose to stay in cabin first four days of honeymoon."[71] Once again, Cammie urged Carrie to come to Melrose, promising quiet, noting that Mary Belle had returned to Shreveport and that Lyle, too, was silent.

In July 1926, *Harper's* published another of Ada's stories. Titled "Maudie," it was a different kind of work for her, depicting a dowdy and unattractive character named Maudie Turner who sells "greasy cosmetics" door to door. Ada wrote in a July letter to Cammie that the story had received a glowing review from Lyle: "Have had such a wonderful letter from Lyle recently concerning 'Maudie.' He likes it better than anything I've done—says it is more 'expert,' a better piece of work. This pleases me very much for I feel that way about it myself, and so did *Harper's*." Carrie was unsure; in April 1926, although she had not yet read the full story, she had heard about it and told Cammie, "I can't imagine what 'Maudie' is like—it doesn't sound 'Ada-ish' enough."[72] *Harper's* obviously agreed with Lyle, paying four hundred dollars for the story.

Also in July, Ada, John, and David took a short trip south along Bayou Teche, with an overnight stop in New Orleans. Ada was disappointed not to see Lyle, who was working, but enjoyed her trip nonetheless and stopped by Briarwood for an hour on the way home. "Carrie is as elusive and hard to catch as one of her squirrels!" Ada exclaimed, and like Cammie, she fretted about Carrie's plans to once again return to teaching in the fall. "I do hope she won't teach next year. Let's do our best to stop her from it," she implored. "You, I'm sure, would have some influence with her."[73] Teaching school drained Carrie's time and energy, leaving little for her writing. Though she continued to share her short stories, poems, and ideas with the Melrose group, teaching full-time made working on them slow and difficult. She did receive a twenty-dollar check from *Holland's Magazine* that summer for a profile of Ada entitled "New Voices from the Old South."[74]

As it turned out, Carrie did return to teaching in the fall of 1926 and at the same time turned down a job opportunity with the forestry department in Mississippi that would have liberated her from the classroom. The work offer, with its regular paycheck, would have been pure salvation for her, but it would also have taken her away from Louisiana, Briarwood, and Melrose. Carrie simply did not want to go and refused the job. She and Cammie discussed the position in letters back and forth: "What will happen next?! Got an offer to do state forestry work in Mississippi, like I did in La. The State Forester says he *must* have me— doesn't know where to get anybody else, and will give me almost any salary I ask! Can you beat it? And I don't want to go! Want to stay in my woods and write and paint. Am I crazy? Please tell me."[75] Never one to withhold her opinions or to limit Carrie's opportunities, Cammie must have advised her to take the job because in a later letter, Carrie wrote: "Don't you want me to stay closer than Miss[issippi]? I did not accept the place. I have just one life to live and I want to spend the rest of this one where I am happy, and with people I love. It was a lovely offer, but it was *public* work and I have already given the best years of my life to the public—the rest I want to live as privately as possible. I'll stay here and manage to pull through somehow. Am sorry you are disappointed."[76] There is no doubt that Cammie recognized Carrie's talents and wanted success for her, but she certainly would have missed her had Carrie actually gone.

Ada's star continued to rise with two more published stories. She received four hundred dollars from *Harper's* for "Singing Woman," which is set in the Isle Brevelle community and concerns two mulatto women, Henriette and Josephine, who are the only two singing women left on Isle Brevelle. A singing woman was a "professional mourner" and was "as necessary as a priest" for a proper burial. It becomes apparent, however, that the mourning ritual is as anachronistic as the old dirt road that runs through the story, enacting a theme of protest against too much progress that comes too fast. "The Old One" is also a mulatto story; "I had old Madame Aubert Roque in mind when I wrote it," Ada explained.[77] An elderly woman, Nicolette, and her grandson, Balthazar, live contentedly on Isle Brevelle until Balthazar marries Rose, who is an outsider. A power struggle between Rose and Nicolette ensues, and as in "Singing Woman," a theme of the old versus the new evolves. In the midst of her success, Ada longed for Melrose, telling Cammie, "I wish it were possible for me to get away and run down to the cabin for a few days." Her best stories centered around the Cane River and its people, and as long as she held her close connection with Cammie and Melrose, she wrote successful stories.

Lyle returned to New York that fall, and in October he finally left his *Times-Picayune* job in order to concentrate on his fiction. His sporadic silences, though typical of him, continued to puzzle Cammie and the Melrose group. It is not that he lived two separate lives exactly, but he did have friends and a busy calendar outside of Melrose. It was necessary for him to return to civilization, as he called it, to write stories or articles that would help pay his bills. That August, Ada had received a "lovely" letter from him and told Cammie: "What a person he is! I told you—didn't I?—that he is the kind you must accept, first as he is. I sometimes feel that if you try to grapple him too hard with hooks of steel he is gone! Hooks of steel always rather frightened me too, I must admit!"[78] Carrie had similar thoughts in November, telling Cammie, "I suppose we just have to accept Lyle as he is—mysterious and all!"[79] She, more than the others, may have understood Lyle's absences from Melrose, as she had to earn her living with lectures and speaking tours. Ada, on the other hand, did not have to publish to survive financially.

Throughout her life, Cammie Henry continued to collect what she called "congenial souls" with whom she would frequently correspond and who would come stay at Melrose. Some people were simply tourists, but others had an affinity for Cammie's interests, and she stayed in close contact with them. Arthur Babb from Dallas, Texas, was one with whom she stayed in close touch. Babb came to Louisiana in April 1926 as a representative of the Texas-Pacific railroad for the purpose of building a depot in Plaquemine, and later that year he traveled to Natchitoches to build the depot located at what is now Sixth Street and Martin Luther King Drive. During his travels, Babb would seek out interesting people, sites, and stories; he took photographs with his Kodak, and he kept a diary that he called his "Sketchbook." One of those people he had heard about and wanted to meet while in Natchitoches was Cammie Henry. Dr. and Mrs. Grant, of Plaquemine, had encouraged him to seek out "Aunt Cammie," and Babb had also read Mary Belle McKellar's article about Melrose in the *Times-Picayune*, so his interest was sparked. Though Cammie's scrapbooks and historical collection initially lured him to Melrose, it was the dynamic woman herself that kept pulling him back there.

Babb was a little nervous about meeting Cammie, noting, "I naturally felt anxious to meet her . . . I began to draw up and feel little" in comparison to great names like Henry E. Chambers, Edith Fairfax Davenport, and Ada Jack Carver. He arrived in Natchitoches in November 1926 and was impressed with the history he found there. He spent his first day walking around, taking pictures of old homes, including the old Carver home, and talking to some of the locals. He also walked along the Cane River and took photographs both inside and outside of the art colony cabin. He was fortunate enough to find Gladys Breazeale working there and discussed art with her for a while, although he admitted he was no expert.[80]

A day or two later, Babb and a friend, J. G. Haupt, had lunch in town and then climbed into a Ford sedan for a drive through the Cane River area toward Melrose; this was an opportune development, as Babb had left his papers of introduction from Dr. Grant back in town, and now Haupt would be his passport into the Land of the Lotus. After winding down along the Cane River admiring the fine homes, the scenery, and the largest pecan trees Babb had ever seen, they finally arrived at Melrose, where they found Cammie working in the yard. She had "a bucket in one hand and a hoe in the other." She set the bucket down, shook hands with Babb, and then promptly scolded him when he snapped her photograph.[81] She gave them the customary tour through the gardens, and then, as it was getting late, the two men headed back to town but not before Babb had secured a precious invitation to return, which he did many times over the years.

It took Babb only two days before making his second visit to Melrose; he noted the date in his Sketchbook as Saturday, December 4, 1926. He and Haupt took the train out of Natchitoches to Fern, which is about one and a half miles from Melrose. They then walked along the road until someone driving by offered them a ride to the Melrose store, where they asked for Mrs. Henry.

At the main house, Cammie greeted her visitors and then led them up the exterior stairs and into the living room, where they found Leudivine reading a book. Babb was charmed with Mrs. Garrett, pronouncing her "one of the most gracious old ladies that I ever met."[82] This time Babb had his letter of introduction from Dr. Grant, which he presented to Cammie. She led him through a quick tour of the home and then to the rear porch, where he found the scrapbooks in tall, glass-fronted bookcases. "I was dazed," he said. "There was much more than I could read in a year."

Babb perused a couple of scrapbooks on Natchitoches, and before long, it was time for coffee, the strong, black drip coffee Melrose was famous for serving. He spent the rest of the morning paging through one scrapbook and then another, totally engrossed, until Cammie rang the plantation bell for lunch. Babb admitted to being almost overwhelmed by the volume of information in the scrapbooks; he wanted to make the most of his time, yet it was simply impossible to see everything. After their lunch, he and Haupt walked through the gardens, with Haupt pointing out the spinning wheels where Cammie worked, Lyle's cabin and its history as a slave hospital, the best places to sit and paint or read, and then the old river, which Babb declared to be "one of the most beautiful lakes that your eyes ever looked upon."[83] He was bewitched by the mystique of Melrose.

Babb had his Kodak with him and snapped one photo after another of cabins, the upper back gallery of the big house, with its shelves and shelves of scrapbooks, various Melrose employees, and the river. He and Cammie continued their friendship long after 1926; they corresponded for the rest of her life, and for many years he sent Cammie a Christmas turkey, at least once with the admonition not to kill it "until December 24."[84] As for Babb's Sketchbook, the original diary was discovered in the Cammie Henry collection some forty-five years after his death, and it remains there today, one more example of the many treasures to be found in Cammie's lifelong compilation of material. In 1996 Neill Cameron edited and published *My Sketchbook*, after securing permission from Babb's family; the edition contains all of the photos Babb took at Melrose.

Melrose was still a working plantation and thus a very busy place. J.H. expanded the pecan orchards, and he always had an eye on the cotton market, which was volatile enough to keep anyone up at night. The workers had been picking cotton for months when the gin caught fire in December 1926, destroying not only the gin but also part of the plantation's profit for the year. In a letter to Carrie, Cammie told her their gin "burned to the ground—30 bales— one seed house—no insurance—bad—could be worse—just so God spares J.H. I'll not murmur."[85] Cotton prices were pretty fair in 1926, but no price on a bale of cotton was worth more than J.H. to Cammie.

Stuck in New York working, Lyle especially wanted to be at Melrose that Christmas, as the city was making him homesick. "I'd give anything I've got to be able to spend Christmas in the cabin—near the people I love, but, I find myself far from everything that means anything to me."[86] Cammie sent him some Melrose pecans and a long letter, which only added to his longing for the South. It would ultimately be the 1927 Mississippi River flood that brought him back.

The first weeks of the "the year of the Big Water," as some referred to it, were busy at Melrose, with the wedding of Cammie's son Joe and Eugenia Cherry in February. Carrie accepted another job with the forestry department, this time as supervisor of forestry education in the Department of Conservation. That spring Cammie and friend Lillian Trichel accompanied Carrie on a trip to South Louisiana for the state conservation department. Lillian was Ada's cousin and had been Cammie's friend for several years. They traveled aboard a little cabin cruiser from New Orleans to the Russell Sage Foundation's wild bird sanctuary on Marsh Island.

Cammie was very nervous about leaving her mother at Melrose, as Leudivine had been ill, but there were plenty of hands on deck to take care of her, so Cammie went. She second-guessed her decision up until the last possible moment to turn back, which was at Morgan City, nearly two hundred miles from home. Once she made up her mind to continue the trip, though, she set her anxiety aside. Carrie was tasked with taking pictures of wild birds for the conservation department. It ended up being "a wild goose trip," as Lillian later termed it, because the birds had flown north before the group arrived.

As it turned out, it was not a wild goose trip at all because along the banks of the canal Carrie spotted some coveted wild irises. Lillian said there were beautiful blue, purple, and white irises growing "in the black water under the moss-draped trees." At Belle Isle the boat docked, and the ladies were able to pull up some roots that had stems close to six feet tall.

Carrie left Cammie and Lillian at New Orleans, and once Cammie had phoned Melrose to assure herself all was well with Leudivine, they got in their car and headed back down to find the wild white irises they had seen from the boat but had been unable to reach. When they finally found them, Lillian pulled the car to a stop, and Cammie asked, "Lillian, can you swim?" "Not much," answered Lillian.

In Lillian's written account of this adventure, she noted: "Miss Cammie thought a moment, then said: 'You can drive and I can't; if I get snake bit you can get me to a doctor. If you get snake bit, you will die right here. I will go in.'" And in she went. Lillian wrote: "Miss Cammie removed her white shirt waist and black skirt, folded them neatly on the grass. Her shoes and stockings were laid beside the skirt and her underskirt and she bravely waded down the bank knee deep— more than knee deep—she gathered four iris stems together to pull up four roots; as she lifted them they came up like the spokes of a wagon wheel—suddenly I was knocked back up the bank and Miss Cammie practically flew over me in one magnificent leap. She had pulled up some kind of animal or fish in the net made by the iris roots; I have always thought it was a sea otter, she could only say it grinned at her with a red mouth and many teeth."[87] Both women dissolved into laughter and fell down on the grass. Not to be outdone, Cammie waded back into the water for her iris and came back out with four roots. She took them home and planted one of them at Melrose, where it thrived, of course.

Lyle's homesickness in New York prompted Cammie to encourage him to re- turn to Melrose in the new year to write. According to biographer Chance Harvey, Lyle was determined to sell himself as a creative writer in New York, and while he did have some success with "The Centaur Plays Croquet" and "Cane River," he had not achieved the creative output he desired. Harvey explains that Lyle spent most of his hours before the typewriter revising stories he had actu- ally written at Melrose, and in fact, both of those stories are clearly influenced by his time on the Cane River. New York did not seem to provide the inspiration he had been looking for.[88]

In May, Lyle at last returned to the South to cover the Mississippi River flood in a series of articles for *Century Magazine,* a national publication featuring jour- nalistic articles, editorials, and fiction. He gathered material for what would eventually be his first book, *Father Mississippi,* which told the story of the river and its history, and offered a plea for help against the flooding. The rains that roiled the Mississippi over the levees also wreaked havoc at Melrose. As early as April, Cammie noted the flooding in a letter to Carrie: "Water getting us—corn land under and 100 pecan trees—15 ft. of water."[89]

From his position on one of the rescue boats in South Louisiana, Lyle wit- nessed the devastation firsthand. He snapped photos of flooded lands, steam- boats crowded with desperate people, levees with stranded animals, buildings with water up to the rooflines, and people standing on top of their homes waiting for help. Copies of these photos are in Cammie's scrapbooks, one tragic scene after another: a man in a boat, poling his way through the water; a house with furniture and household possessions on the roof.[90]

Lyle attempted to keep a journal through these days. His 1927 diary is a thin volume that he could easily carry in a shirt pocket; the date is stamped on the front cover in gold, and inside the front cover is written his name and a request that the volume be returned to the *Times-Picayune* if lost. His entries are sporadic, written mostly in pencil, and there are random names and phone numbers jotted in the margins. On May 12 he recorded that he left New York "for flooded area to get material," arriving in New Orleans on May 14. He was off to Baton Rouge the next day, and by Monday he was aboard the *Saulsee* and launched into the flood. While the world watched Charles Lindbergh's transatlantic flight, Lyle Saxon documented the unfolding tragedy of the flood and its impact on the residents of the Mississippi Valley. He recorded nothing except where he was for the next few days—"Bayou des Glaises" and "Melville Crevasse."[91] "The flood is so vast," he wrote to Sherwood Anderson, "that I can't describe it at all—all the same everywhere, ruin and desolation, dead cows, dead horses, dead chickens, dogs marooned in trees, starving."[92]

In June he took his material and headed to Melrose, finally, to work. He was feeling the pressure of his deadlines as he wrote on June 5 to Anderson: "This flood business is driving me crazy. I've signed up for four articles for *Century* magazine . . . and for a book on the Mississippi River, old steamboat days and all that kind of thing, and dwelling at length on the flood of this year. I signed a contract to deliver the manuscript to them by August 15—and that means writing 2,000 words or more a day . . . it's worrying me sick."[93]

He stayed in his cabin at Melrose and worked most of the summer. In the evenings he often sat on the sleeping porch at the big house with Cammie and Leudivine, where they read to each other. When necessary, he would take the train to Baton Rouge to see his aunts or take care of business. On August 8 he scrawled a note to Cammie on the back of a hotel laundry ticket informing her that he was heading to New York; he still had not "finished the damned book" and was taking some of her scrapbooks with him.[94] A week later, he noted in his diary that the book was finished, and on August 16 he wrote that he had "delivered the manuscript of *Father Mississippi* to Century." The book was published at the end of October to mostly excellent reviews across the country, but to Lyle it all seemed rather surreal. He told his Aunt Elizabeth: "It doesn't seem possible that all those messy pages that I sweated over last summer should evolve at last into anything as tangible as a book—and people should actually pay good money for it . . . it is strange, but all this fuss means nothing to me . . . I can't get one thrill out of it—like a dead man."[95]

Cammie could barely contain her elation over Lyle's success. When the book came out, she declared, "Lyle's fortune done made!"[96] Critic Cleveland B. Chase of the *New York Times,* however, was less than impressed. He criticized Saxon's "extensive citations from well-known passages in the reports of early explorers," which were "already easily accessible and add little to the interest of the present volume," as well as his heavy use of previously unpublished diaries that "are scarcely of a nature to fit into the general plan of the book." Even more scathing, Chase said that "one gets the unfortunate impression that the author, in his haste to prepare one of the first volumes to appear after the flood, laid hand upon whatever material concerning the Mississippi Valley he could gather without great effort."[97] Incensed, Cammie wrote, "Couldn't have been done otherwise," in the margin next to the awful review she had pasted into a scrapbook. Lyle's friend, writer Josiah Titzell, was disgusted with Chase's review as well. "This is a stupid review," he said. "Every other review in the U.S. praised the book. This man has no literary sense whatsoever! A dull attitude to take [about] a book which every other reviewer praises."[98]

The reviews did look very good otherwise; friends from all over the country sent those they came across to Cammie. Mrs. J. W. McCook mailed a copy of Bernard DeVoto's February 1928 *Saturday Evening Post* review, in which he wrote: "One must insist on its importance. It is a book that will create lesser ones." He goes on to call the work "exquisitely done."[99] Some critiques were mixed and seemed to focus more on "debating the proposed flood legislation before Congress than on evaluating Saxon's book."[100] *Father Mississippi* has stood the test of time, and seemingly every book about the great flood since then makes reference to or quotes from Lyle's book. There have been more scholarly works, to be sure, but Saxon's book served the immediate purpose for which it was intended. He has been credited with coining the designation *Mississippi Valley* as an area "not connected by growing seasons as was the Confederacy, but by its vulnerability to the river."[101] In his reporting, Lyle was equally sympathetic to both black and white refugees and sharply critical of what was seen by many as government overreach in handling the river.

Given the timeline of events and Lyle's short deadline, the book had to be composed hurriedly. No doubt he relied heavily on Cammie's scrapbooks for source material, which he combined with his own firsthand observations and opinions. He had been in New York when the flood started in the spring, was in Louisiana in early May, and was in the cabin ready to write by mid-June. The book was at the publishers in mid-August, a mere two months later, but it stands up even today due to Saxon's storytelling skills and his adroit newspaperman's perspective. *Father Mississippi* is less a dense, research-laden tome and more a story one might hear sitting at a great uncle's feet by the fire. To Lyle the book must have seemed like simply an extension of his newspaper work. What he still really wanted to do was to write good, marketable fiction, and in his letters he seemed much more interested in the success of his stories "Cane River" (1926), "The Long Furrow" (1927), and "The Centaur Plays Croquet" (1927).

Reflective about his work in a 1930 unsigned interview with the *Baton Rouge State-Times*, he explained with regard to *Father Mississippi* that the introductory chapters of a young boy's impressions of the great river—chapters some took quite literally—do not represent his personal experience: "It is not my own story necessarily, but any boy's story of the horror of a crevasse destroying everything he held dear." Following those initial chapters, he gave the history of the river and then finally an account of the 1927 floods, which, he said, "I saw at first-hand." "In *Father Mississippi* I tried to show the eternal danger of the river, hoping that the book might help arouse Congress to give us proper flood control in the lower valley."[102]

While Lyle monitored the flood and worked on *Father Mississippi*, Carrie had begun reworking "The Sandhill Tales," which Lyle promised to read and critique for her. Ada had been urging her for years to complete the stories, and Cammie also encouraged this endeavor, telling Carrie, as always, to come to Melrose to write. "The Sand Hill Tales" is a collection of stories that Carrie drafted off and on for at least three decades. She described them as "stories of the poor whites of the backwoods of Louisiana, Alabama, and Georgia." They are "such quiet little things," quite different from the type of stories that Ada Jack Carver wrote. There is, she said, "no punch to hit an editor between the eyes." Ada's stories are driven more by narrative and atmosphere, while Carrie's "Sand Hill Tales" are more dialogue oriented. Editors apparently agreed with her, as there is a stack of rejection letters that Carrie saved along with her stories. A New York literary agent, Robert Thomas Hardy, rejected "Possum," a "Negro story" representative of a popular genre at the time. Hardy's rejection letter to Carrie declared the story "trite" and without "enough of a point." He then advised her to "not write this type" of story.[103]

A more typical example of the stories from the "Sand Hill Tales" is "Molly's Children," which tells of a married couple working hard to get by with six kids who all need food and shoes for school. The husband, Doc, saves a little money working a side job for the purpose of fulfilling a lifelong dream of seeing the Louisiana State Fair in Shreveport. Meanwhile, Cousin Molly dies, leaving two hungry children of her own. Doc and his wife take in the two orphans, thereby forgoing their dream of seeing the giraffes and monkeys at the state fair. It is, as Carrie said, a quiet little story about good people working hard and doing the right thing. Carrie's "Sand Hill Tales" and many of her other stories are in the same general style and are more complex than they appear, often threaded with symbolism, imagery, and moral lessons.

It is interesting that there never seemed to be any kind of envy, jealousy, or competition between the Melrose writers. Carrie and Ada Jack celebrated each of Lyle's successes and encouraged his desire to write fiction; all of Ada's successes were met with support and celebration as well. If anyone ever had a reason to feel any envy at all, it would have been Carrie Dormon, but there is no evidence that this was ever the case. Carrie did publish a plethora of nonfiction articles, but in a 1947 letter to friend and writer Ann Titzell, she noted that she had "never sold one word of fiction": "[I] have been selling articles since 1922. And I have never stopped writing."[104] Carrie seemed to believe that all she needed was a good agent, and she asked her friends Josiah and Ann Titzell to recommend one, noting that Mr. Uzzell "simply made Ada Jack Carver." "Not that she did not have unusual gifts," said Carrie, "but she herself told me that he made her do 'Redbone' over completely before *Harper's* took it." Carrie continued to send off her stories on her own without an agent.

Cammie encouraged Carrie through a series of several letters to send her stories to Lyle. "Be ready to send all materials by Aug. 20," she insisted, giving Carrie a deadline in 1927. "It's now or never—why miss a golden opportunity, then rig your ship to come work in the cabin."[105] Five days later, Cammie wrote her another letter. "Send your work to him this month—don't you dare fail," she urged.

Carrie did get her work together for Lyle by Cammie's deadline; she sent some poems, short stories, and blueprints. By the end of the second week of September, she told Cammie that she had received "the nicest letter from Lyle," who "had just got my stuff—hadn't taken it to anybody."[106] When he responded to Carrie, Lyle told her, "You certainly have done something fine," and said that he found "The Sand Hill Tales" especially moving, stirring his emotions "just as they did when I first heard you read" from them.[107] He planned to take the stories to Hewitt H. Howland, the editor of the *Century Magazine*, but cautioned it would be "a week or so" before he would hear anything. Carrie was grateful but anxious: "I try not to hope but just can't help it!" The stories, over twenty of them, were never published.

Carrie continued working on her nonfiction, and in February 1928 the Louisiana Conservation Department published a booklet she had penned, *Forest Trees of Louisiana and How to Know Them*, which was distributed free to teachers across the state. She was also getting closer to her dream of creating a national forest in Louisiana, a project that had been a decade in the making. That month the National Forest Reservation Commission approved the future acquisition of 9.6 million acres of land for preservation, including a portion of virgin pine forest that would later become the Kisatchie National Forest. In March 1929 the *Times-Picayune* reported the purchase of 365,000 acres of forestlands in Grant, Natchitoches, and Rapides Parishes. Paul Wooten's article mistakenly identified Carrie as "Katherine" Dormon (a gaffe that Cammie corrected in the margin of the scrapbook where she pasted the clipping) but correctly pointed out that Carrie had been working tirelessly since 1923 for this moment. In July 1930 the National Forest Reservation Commission decided to purchase 425,000 acres of cutover lands in Louisiana for reforestation. Elated, Cammie declared this was "due wholly to Carrie: her monument."[108] Ultimately, this would be her legacy. As much as Dormon aspired to write and publish short stories, her expertise had been recognized in her many published books and articles on Louisiana plants, flowers, and conservation.

ammie stayed busy in the spring of 1928 restoring cabins, having one of them covered with cypress boards. She was also filling scrapbooks with *Father Mississippi* reviews and had a massive amount of daily correspondence to keep up with. She spent most mornings and evenings in the gardens, supervising and directing work there. In an undated piece that Carrie wrote (possibly as part of the "Melrose Stories"), she described Cammie as being "in all places at once." She depicted Melrose as a beehive of activity as Cammie directed the hands at their weeding, manure spreading, and planting. Carrie characterized Cammie's green thumb as nothing short of miraculous: "At the birth of her second son, she planted a live oak. In no time at all it was a spreading tree, and now provides a circle of welcome shade more than a hundred feet in diameter." Carrie also described the hazards of gardening at Melrose: "The inexhaustible supply of soft water from Cane River Lake was one of the secrets of her success with plants. And she never trusted anyone entirely with this watering, but supervised it herself. In summer she kept the ground so wet she had to wear galoshes in the garden. And these were usually dropped in a corner of the cement porch, which was flush with the ground. One morning she found it impossible to force her foot into one of them. She took a look, and there was a small moccasin coiled comfortably inside!"[109]

Throughout their lifelong friendship, Carrie collaborated frequently with Cammie on the Melrose gardens. She would send plants cultivated at Briarwood to Cammie, and Cammie would likewise send plants not only to Carrie but to almost all of her friends. In most letters to Carrie, Cammie reported on what was currently in bloom at Melrose. A March 1926 letter recorded the vigorous growth of various trees and shrubs, with buds and leaves enclosed for Carrie's inspection. Virtually everything Cammie planted survived and thrived. If Carrie suggested that Cammie was planting something too close to another plant, Cammie would point out that was how they grew in the wild, and indeed, they flourished under her care.

Lyle was at Melrose that spring, but by the summer he was back in Baton Rouge, working on *Fabulous New Orleans,* which did not appear to be going any better or giving him any more satisfaction than *Father Mississippi. Fabulous New Orleans* is a series of impressions of the city that captures its history and culture—beginning with Mardi Gras, of course, and a young boy's impressions the first time he is taken to the parades, masked, by a family servant. There are also sections on the quadroon balls and Marie Laveau and a sort of narrated tour through the Vieux Carré. When he wrote to Carrie in June, Lyle lamented that the book "won't end. It won't even come near ending," and was due in two weeks to his publisher.[110] The book came in at about one hundred pages less than *Father Mississippi* but was filled with delightfully rich material drawn from his own experiences, history books, and Cammie's scrapbooks. Lyle now had two successful nonfiction books to his credit, and his fame continued to rise.

In the summer Mary Daggett Lake, a writer for the *Fort Worth Star-Telegram* and president of the Texas Federation of Garden Clubs, visited Melrose to gather information for an article. In September she published an account of her trip, titled "A Pilgrimage to the Heart of Old Louisiana," in which she made particular note of one of Cammie's prized possessions: the McAlpin stencil that hung above the living room door.[111] Cammie acquired the famous copper stencil on one of her explorations of Isle Brevelle: "I found this copper stencil in [the] garret of Marco house on lower Cane . . . this side of Colfax," she wrote in one of her scrapbooks.[112] The old Marco house, built in the 1820s or '30s, had been slated for destruction, and Cammie was there to salvage what she could of the once beautiful home. Along with the stencil, she also purchased a large fan light that she installed at Melrose.

Stencils of this type were used in marking bales of cotton, but copper stencils, as François Mignon later explained, were "a hallmark of affluence and distinction."[113] Robert McAlpin had a plantation in Natchitoches Parish and was allegedly the model for Simon Legree in the famous Harriet Beecher Stowe novel *Uncle Tom's Cabin*. Many believe that Stowe visited McAlpin plantation for inspiration: Cammie noted that "Mrs. Stowe staid [*sic*] at Robert McAlpin's house when she came to watch and gathered data for her book. I have often heard Dr. Scruggs recall Mrs. Stowe's visit to his father's house."[114] The stencil was one of Cammie's favorite treasures and was one of the things she always showed to visitors.

The year 1929 was notable not only for the stock market crash but also for the death of Cammie's friend and correspondent Henry E. Chambers, who died of a stroke in March. Cammie simply noted the date of his death in her scrapbook alongside one of the last letters she had received from Chambers. In October the stock market crash created panic on Wall Street and across America. While there is no record of Cammie's immediate reaction, it is likely that J.H. discussed the situation in correspondence with his brother Stephen more so than with his mother. The financial crisis would eventually reach even Cammie Henry's rural plantation on the Cane River.

The Christmas of 1929 was a white one at Melrose, which was rather unusual for Louisiana. Ada visited during Christmas week, spending two days and one night on the plantation.[115] In a possible nod of acknowledgment that she had not often been to Melrose recently, Cammie categorized her visit as "a picking up of dropped stitches." Over the previous few years, Ada had been withdrawing more and more from the Melrose group; it had been two years since her last story appeared in *Harper's*. In the fall of 1929, Cammie gathered typescripts of Ada's stories and had them bound in a volume for Ada's mother. Mrs. Carver's note of thanks refers to "the compiled material which is so precious to me and [is] not in such available form! . . . I can never tell you how I value it—surely it was a most beautiful tribute to a friend to have done it so carefully . . . when I told little Ada over the phone she was deeply touched by it."[116] It was a simple act of kindness that was so typical of Cammie Henry.

At Briarwood, Carrie settled in with pen and paper to record the snowy scene outside her window for Cammie: "If only you could see Briarwood this afternoon! Snowflakes whirling, pines and cedar already decked out like fairy Christmas trees! The pond is a black mirror with snowy hillsides reflected in it. And the cabin looks more than ever as if it just grew up out of the ground . . . I've been here a good many years, but I never saw anything like this before! The snow is fully 8 inches deep on the level and some drifts are 2 ft. deep! . . . The only signs of life are bird and rabbit tracks."[117]

Lyle spent the last six months of 1929 in New York. His third nonfiction book, *Old Louisiana*, came out that October and was receiving favorable reviews. Gilbert W. Mead, reviewing the book for the *Birmingham News*, wrote, "Mr. Saxon has reconstructed the old life with a patient fidelity which proves his deep love for the scenes he describes."[118]

The 1920s had been fruitful for each member of the Melrose circle. Perhaps the biggest success, though, was Cammie Henry's. She had entered the decade as a new widow with eight children to raise and a working plantation to manage.

By its end, her children were either grown or successfully settled into school, with the youngest, Cammie Jr., graduated at age fourteen from the Louisiana Normal School and soon headed to Winston-Salem Academy in North Carolina, the alma mater of her grandmother Leudivine. Melrose was thriving under the management of J.H., and Cammie could see her life's path as patron and muse to writers and artists, archivist of her own historical library, and matriarch of the Henry family. Her mission was clearer to her than ever.

Cammie Garrett, age 18. *Inscribed on reverse:* "To my darling Mother, Christmas Eve, 1889."

(Joseph M. Henry Collection, folder F11A1, Cammie G. Henry Research Center [CGHRC], Watson Memorial Library, Northwestern State University of Louisiana)

Louisiana State Normal School graduating class, with Cammie Garrett
(*seated on ground near center, face in profile*), 1892.

(Melrose Collection, SB 74, 43, CGHRC)

The main house at Melrose Plantation in 1899,
when Cammie and John Henry moved there.

(Melrose Collection, SB 79, 10, CGHRC)

Cammie Henry feeding chickens at Melrose in 1908.

(Melrose Collection, SB 79, 8, CGHRC)

In 1909, Cammie and John had six children: Stephen G. Henry (*standing in back*), age 15; John Hampton Henry Jr. (J.H.), age 12; Isaac Erwin Henry, age 10; Joseph Marion Henry, age 7; Daniel Scarborough Henry, age 11 months; and Payne Walmsley Henry, age 3. They would have one more son and a daughter before 1915.

(Melrose Collection, SB 196, 47, CGHRC)

Melrose blacksmith shop, 1909. Two unidentified men work on a plow. The African House (*immediate left*) and Yucca (*far left*) are visible. It appears to be laundry day.

(Melrose Collection, SB 68, 105, CGHRC)

Cammie Henry's library at Melrose, 1916.
(Melrose Collection, SB 190, 6, CGHRC)

Front of the main house, Melrose Plantation, 1920s.

(Melrose Collection, SB 83, 1, CGHRC)

Israel Sudduth in front of the main house, 1920. Cammie's notation: "Aunt Jane's husband—he was raised on Melrose." (Melrose Collection, SB 68, 106, CGHRC)

St. Augustine Catholic Church and rectory (*left*), Isle Brevelle, ca. 1922.
(Melrose Collection, SB 71, 28, CGHRC)

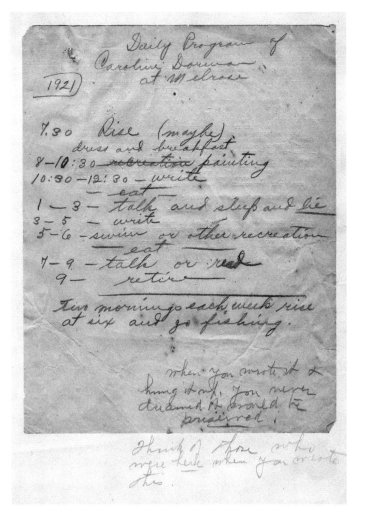

Caroline Dormon's handwritten daily schedule during a visit to Melrose, 1921.
In the bottom right corner Cammie later wrote: "When you wrote it &
hung it up, you never dreamed it would be *preserved*."

(Melrose Collection, SB 83, 23, CGHRC)

Caroline Dormon's home, Briarwood, after a snowfall, February 1923.

(Melrose Collection, SB 83, 12, CGHRC)

Lyle Saxon in the cabin called Yucca, July 1924.

(Melrose Collection, SB 220, 2, CGHRC)

Caroline Dormon and Ada Jack Carver at Melrose, probably 1925 or 1926.

(Caroline Dormon Collection, folder 566, CGHRC)

Lyle Saxon, Alberta Kinsey, and Ada Jack Carver at Yucca House, probably 1925 or 1926.

(Melrose Collection, SB 83, 63, CGHRC)

Cammie Henry at a spinning wheel in front of one of the plantation cabins, ca. 1927.

(Melrose Collection, SB 220, 2, CGHRC)

Cammie Henry and Clement Claiborne making shuck bottoms for chairs, 1929.

(Melrose Collection, SB 244, 56, CGHRC)

The McAlpin copper cotton stencil of which Cammie was so proud.

(Melrose Collection, SB 68, 47, CGHRC)

The African House when it was used as a barn, 1929.

(Melrose Collection, SB 79, 83, CGHRC)

Cammie Henry and her driver, Fuggerboo, in the future
Kisatchie National Forest, 1929. Photo by Caroline Dormon.

(Melrose Collection, SB 72, 116, CGHRC)

Christmas 1929 at the front entrance to the main house, Melrose Plantation.
From left: Mary (house worker), R.E. (odd jobs), Fuggerboo (driver), Messaline (cook),
and Bertha (kitchen worker). Cammie did not include their last names
in the caption but did indicate their work roles.
(Melrose Collection, SB 203, 51, CGHRC)

Cammie Henry and her mother, Leudivine Garrett,
at the front entrance of the main house, n.d.

(Melrose Collection, SB 79, 8, CGHRC)

Cammie Jr., age 16, at a druggist convention in Dallas, 1931.

(Melrose Collection, SB 203, 142, CGHRC)

Lyle Saxon and Cammie Henry at Melrose, n.d.

(Melrose Collection, bound vol. 182, 108, CGHRC)

Melrose Plantation, February 1940.

(Photo by Lester Jones, Library of Congress Prints & Photographs Division, HABS LA, 35-MELRO, 1-1)

Front entrance of the main house, February 1940.
Cammie's oil jars can be seen on the porch.

(Photo by Lester Jones, Library of Congress Prints & Photographs Division,
HABS LA, 35-MELRO, 1-3)

Rear of main house, February 1940. The back wing was added some time after John H. Henry's death in 1918. Its upper story included Cammie's bedroom and a screened porch (*right*), where she kept looms and spinning wheels; its lower floor was the kitchen.

(Photo by Lester Jones, Library of Congress Prints & Photographs Division, HABS LA, 35-MELRO, 1-2)

Frenchie's store, bar, and gas station near Melrose Plantation, 1940.

(Photo by Marion Post Wolcott, Library of Congress
Prints & Photographs Division, LC-USF34-054355-D)

The Melrose plantation store, 1940.

(Photo by Marion Post Wolcott, Library of Congress
Prints & Photographs Division, LC-USF34-054608-D)

Tenant house on Melrose Plantation,
with cotton growing right up to the front door, 1940.

(Photo by Marion Post Wolcott, Library of Congress
Prints & Photographs Division, LC-USF34-054668-D)

Two members of the Roque family near
Melrose Plantation, 1940.

(Photo by Marion Post Wolcott, June 1940, Library of Congress
Prints & Photographs Division, LC-USF3-054681-D)

Geese in a road near Melrose Plantation, 1940. The dirt roads
made Melrose virtually inaccessible in the rainy season.

(Photo by Marion Post Wolcott, Library of Congress
Prints & Photographs Division, LC-USF34-054712-D)

Henry Hertzog and his son Joseph, Melrose Plantation, June 1940.
Hertzog was skilled in threading Cammie's looms and in carpentry and woodworking.
(Photo by Marion Post Wolcott, Library of Congress Prints & Photographs Division,
LC-USF34-054933-E)

Cammie Henry and François Mignon in the front yard of the main house, 1945.
Mignon commented in his journal that this pose was uncharacteristic
of Cammie, who almost never sat still in the garden.

(Photo by Frances Benjamin Johnston, Melrose Collection, folder 84, CGHRC)

Cammie's scrapbooks from the 1930s on rivers and boats of Louisiana, housed in the Melrose Collection, CGHRC.

(Photo by author)

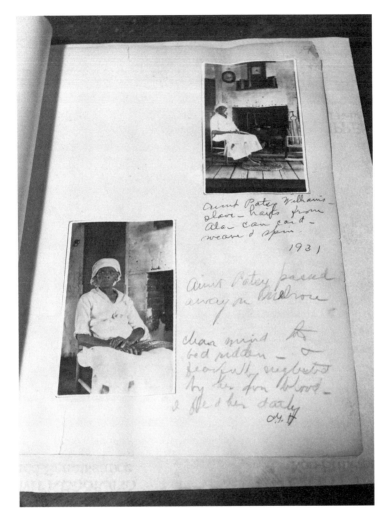

A page in one of Cammie's scrapbooks. *Top photo:* "Aunt Patsy Williams—
[former] slave—hails from Ala[bama]. Can card, weave & spin. 1931." *Bottom photo:*
"Aunt Patsy passed away on Melrose. Clear mind tho bedridden—& fearfully
neglected by her own blood. I fed her daily. C.G.H."

(Melrose Collection, CGHRC)

Cammie Henry's bedroom as preserved today. Her shirtwaist lies on the bed, along with a red-and-white coverlet she wove on a loom.

(Photo by author)

4

A Widening Influence, 1930-1936

The servants with morning black drip coffee tiptoe noiselessly on their rounds.
Even Cane River, as it courses by the plantation, murmurs softly.

—HAROLD M. CASE, "Melrose—Mecca of the Muse—Jewel of Cane River"

When Lyle returned to Melrose in early 1930, he brought with him a contract for his novel, which he had promised to his publisher by May, as well as a contract with Century for his fourth nonfiction book, due August 1.[1] His biographer James Thomas records that the publisher for the novel was Harrison Smith, who gave Lyle a one thousand–dollar advance for the book. Though he really wanted to work on his novel, as of yet he had not even begun it. Lyle is quoted in François Mignon's book, *Plantation Memo,* as saying he first conceived the novel, *Children of Strangers,* in 1925.[2] Biographer Chance Harvey contends that it was "a project that spanned over a decade, and was started as early as 1923."[3] Saxon's published short stories "Cane River" (1926) and "The Long Furrow" (1927) were both really studies for the novel.

Lyle had spent most of the month of January "drinking heavily" in Baton Rouge before going on the wagon at Melrose.[4] In his 1930 diary, he confessed that he was "feeling bad and tired out." Although Prohibition would not be officially repealed until 1933, Saxon and anyone else in New Orleans who wanted to consume alcohol had little trouble obtaining it. Joy Jackson notes that "by 1924 smuggling of liquor in New Orleans had become a large and lucrative illicit operation," and even making home brew was a common practice.[5] Alcohol was available through waiters in restaurants, shopkeepers, taxi drivers, fruit stand operators, and bellhops, depending on who you were and who you knew. Faulk-

ner's biographer Joseph Blotner states that when Faulkner, who roomed for a
time with Saxon in New Orleans in the late 1920s, came to New Orleans in 1925,
he "needed to fear the absence of something to drink no more than the lack of
one good meal a day."[6] This wide availability easily spread to other parishes,
and there were plenty of bootlegging operations in place around Cane River. As
well-known and as popular as Saxon was, he could have obtained whatever alcohol he desired, although he favored absinthe and Sazerac.

Saxon's diary entry for January 21 reads: "Left Baton Rouge for Melrose—
Back at cabin again and glad to be here. Bitterly cold . . . no one here except
Aunt Cammie, little Cammie, Dan, and Mrs. Garrett . . . R.E. and Fugaboo met
me at the train with much giggling." Lyle's presence at the plantation brightened
everyone's mood, and the Melrose employees appear to have been very fond of
Lyle, as he was of them. On at least one occasion, Melrose employee Henry Jack
wrote a letter to Lyle in Baton Rouge expressing his wish that Lyle would return
to the plantation, explaining that Melrose "is so lonesome with-out you."[7]

One method of communication Cammie and Lyle used when he was at Melrose was the "War Baby Express," which was nothing more than using a Melrose hired hand known as "War Baby" to ferry messages from the cabin across
the yard to the big house. One evening, for example, there had been a debate
between Cammie, Leudivine, and Lyle about a word they had come across in
their nightly reading. Upon retiring to his cabin that night, Lyle pulled out his
dictionary and did the research; the next morning he sent a note via the War
Baby Express: "Right as usual Mrs. Henry. There is such a word in the Century
Dictionary. Here it is: ALBATA (from the Latin Alba, meaning white); . . . so you
and Miss Leudivine were right while Dr. George and I were merely ignorant. It
is odd that neither of us ever heard it before. L.S."[8]

War Baby himself provided much mirth at Melrose and was the basis for
more than a few humorous anecdotes that Lyle recorded in letters to his friends.
The chief driver at Melrose was Fuggerboo, but Lyle recounted in a letter to
Carrie the one day War Baby decided to drive: "Instead of telling Fuggerboo to
get the car for young Cammie, [War Baby] decided to back it out of the garage
himself. He succeeded beyond his wildest expectations, backing with the speed
of a hurricane into the store gallery, knocking down a post, busting an axle, and
throwing himself first into the windshield, then bouncing into the back seat."[9]
War Baby was not hurt, and J.H., who witnessed the incident, just laughed so
hard he could not do a thing about it.

Carrie was also able to get down to Melrose in January 1930, spending two days at the end of the month. The trio spent the evenings before the old Ben Franklin stove in Cammie's room, reading to each other from books or from stories they were working on. They spent the crisp, cool days gallivanting up and down Cane River, visiting neighbors. Lyle always made it a point to visit Zeline Roque—an Isle Brevelle neighbor on whom one of his characters in *Children of Strangers* is based—and Madame Aubert Roque when he returned to Melrose. One afternoon they drove to Boyce to see an old house.

Ada's health was troublesome as she fought asthma and malaria off and on for many months. She wrote in the fall of 1929 that she expected to go to New Orleans for an operation on her sinuses that winter and hoped it would finally cure her of "this terrible fatigue that has sapped my strength the last two years."[10] Even though her Melrose visits were seldom, she and Cammie continued to write affectionate letters. At one point, Ada wrote: "The memories of Melrose, and all our good times are always in my thoughts, however. You are so precious to me, and mean so much in my life."[11]

In mid-February 1930 Ada planned a trip to Melrose for a few days, writing that John would drive her to Natchitoches and "if you can send Fugaboo for me then I plan to spend the rest of Friday, Saturday and Sunday with you," adding that she was bringing something to work on. John would pick her up in Natchitoches at the end of the weekend. During this visit, Ada and Lyle spent long evenings talking in the cabin, and in his diary Lyle recorded that he had had "good talks with Ada. Autographed my three books for her."[12] When Ada returned to Minden, she composed a long thank-you letter to Cammie for the visit, which Cammie preserved in the Ada scrapbook; in the margin next to the letter, Cammie wrote, "Written on return from her stay (wkend) at Melrose—it was an unmarred visit."[13]

Cammie's comment is rather cryptic, and Ada's letter gives few clues as to what an "unmarred visit" might entail. Ada referred to the weekend as a "delightful visit—so very satisfactory in so many ways." To Cammie she wrote: "I have thought over it all, gone over it all many, many times a day—This morning a bowl of jonquils on my table brings to me the very essence of Melrose in spring . . . The past three years have been such happy years for me and my own little family. Perhaps—as Lyle suggested, this is the very reason I haven't thought it important or necessary to push my stuff to completion, and to let the world have it (not, however, that the world is clamoring for it!) For you see—it's here, tons of it—so much that I am appalled when I look at it. But anyway, you, as always (and by you, I mean Lyle too) have helped me reconcile myself, adjust myself to the two worlds I live in, both of which, at present, are very real."[14]

In the same letter, Ada also noted that "it's a mistake to let too long a time lapse between seeing one's friends." It seems she was torn between her life as an author and her family. She had not published anything in some time but often mentioned stories on which she was working. Ada still longed for and felt rejuvenated by spending time at Melrose, but something was off.

T he big house was filled with company that spring. In early March, Leudivine wrote to Carrie that there were "so many people about, regular rush of company, one night every bed full. Fred Wilson had to go over to the cabin with Lyle. Miss Lay still here & two other ladies. All got together last night in Cam's room and told ghost stories."[15] Some nights at the gathering in Cammie's room after supper, Lyle would read to them. Cammie's son Joe told Saxon biographer James Thomas that sometimes Melrose seemed like a "big hotel" and that he could recall "when they would cook thirteen chickens for lunch."[16]

In his 1930 diary, Lyle records books he was reading and excursions he and Cammie took throughout the parish. One afternoon in March, they went antique hunting in Cloutierville, about ten miles from Melrose. Sam Lacaze, a friend of Cammie who had a general store there, had a full storeroom and had given Cammie permission to go through it, so she, Lyle, guests Kitty Lay and Janet Field, daughter-in-law Celeste, and Dan's visiting friend, Fred Wilson, all went to investigate; it was well after dark before they returned to Melrose. A couple of days later, Cammie, Lyle, Fred, and Janet returned to Cloutierville, "to the attic of old Menoist house to rummage among old books and rats nests."[17] A disgusted Lyle records, "Didn't write one god-damned line today."

The very next night, the little group returned to the Cloutierville attic armed with "only flashlights," and they "got two sacks of books and a bed, (including canopy), not to mention sundry picture frames and odds & ends, and brought it all ten miles to Melrose in the car with us. Great fun. We had coffee at midnight or nearly that and laughed ourselves sick over our silly venture."[18] Saxon recalled the escapade gleefully in a letter to Cammie Jr. a year later, saying: "I still laugh every time I think of the night that we all went to Cloutierville and went up in the attic there . . . Wasn't that ridiculous? I enjoyed it. I haven't laughed so much since."[19] The whole adventure is illustrative of the extent Cammie would go to acquire antiques she felt had value and must be preserved. The following day they packed a picnic lunch and headed to Briarwood to spend the day with Carrie.

With continued company, the diversions and antics were never ending. One evening Cammie held a spontaneous "quilt show" with Leudivine, who Lyle said was "in great spirits." He noted, however, that not all members of the audience were as enthusiastic: "Fred and I held up quilts. Aunt Cammie gave histories. Miss Lay looked. Janet yawned. J.H. came in briefly. Dan went to town to escape."[20] While these furniture hunting escapades, quilt shows, and picnics were all great fun and characteristic of daily life at Melrose, they did keep Lyle from his objective, which was to finish his novel. He did not seem to have the willpower to refuse the invitations or the activities, and he grew increasingly irritated with himself as the cycle of leaving and then returning to Melrose for the purpose of concentrating on the book repeated itself.

As the spring wore on, Lyle's impatience with himself and with the continuous parade of guests at Melrose become more evident in his diary. He mentioned a guest, "the red-headed woman," who was "beginning to get on my nerves . . . I read and write and avoid the big house as much as possible."[21] Another guest was described as "a wild-eyed old lady who shook hands like a fish, mumbled to herself and speared biscuits with a fork." He would stay at Melrose only a few more weeks, long enough to see an unseasonable snowfall the last week of March that left them all "flabbergasted." Saxon describes the day as "warm and pleasant," but "late in the afternoon it began to snow. Nobody could believe it but there it was. We ran around like mad, covering the tomato plants in the garden—Fred, Aunt Cammie, Carrie" and a host of Melrose employees fighting to protect delicate spring blooms. "By dusk the ground was white. 'The oldest inhabitant' has never known snow so late in the year."

Lyle's 1930 diary is one of a few sources that help give a good glimpse into daily life at Melrose. His entries are brief, usually only a line or two, and he was fairly diligent about writing through the middle of April, when he returned to Baton Rouge. He recorded such mundane things as his trip into Natchitoches for a haircut and what he was reading (in February he was rereading William Faulkner's *Mosquitoes*). This entry is typical: "Drab day—warm and sunny. Gave Massaleen a dollar stipulating that she get drunk on it, which she promised to do . . . Dan and Fred returned late & drunk. Dan lecturing on The Monroe Doctrine at the top of his voice. I could hear him in the cabin. Aunt C luckily slept through it."[22]

We can also see in these entries Lyle's fondness for Fred, who had arrived at Melrose in mid-February and presented Lyle with the gift of a diary. Dan and Fred would come to Lyle's cabin for coffee and conversation, and on one occasion they lured him away from his typewriter to go catch crawfish. Everyone at Melrose enjoyed Fred's company. In December 1930 Cammie received the shocking news that he had died suddenly in New Orleans at the age of twenty-one. Notifying Carrie, she wrote: "You may know—but for fear you don't, I write—we are deeply grieved—Precious Fred Wilson III dropped dead on the st[reet] of N.O. yesterday—we know few particulars—strangers found him—no identification— but K.A. [Kappa Alpha] pin. Frat house immediately notified—boys rushed to scene—Fred was taken to hospital—all in vain—he was gone! Oh, that mother! Her cup is truly full."[23] Carrie, by return mail, wrote: "I am just sick about poor Fred! It is so pitiful. And his poor mother! She has lost her two oldest sons."[24] Lyle took Fred's death quite hard, mentioning him several times in subsequent letters to Carrie.

By the middle of April 1930, Lyle decided he had dallied long enough and not enough work was getting done. He broke the news to Cammie that he was leaving and in his diary wrote, "Why, I wonder, is it so difficult to go?"[25] He headed to Baton Rouge to finish *Lafitte the Pirate* (another book for which he drew heavily from Cammie's scrapbooks) and to attempt progress on the novel. Cammie almost certainly hated to see Lyle leave the plantation, but she had a family Easter gathering to plan. Ada Jack was scheduled to come, but as had become her pattern, she canceled at the last minute through a letter she sent a couple of days before the weekend.

One of the main functions of the Melrose circle was the support and encouragement each member lent to the others, and in the spring of 1930, Lyle chose Carrie Dormon to write a profile of him for *Holland's Magazine*, primarily a regional women's publication featuring both fiction and nonfiction articles. The profile would be similar to the one she had written about Ada years earlier. He had been approached by another writer who wanted to do the piece, but Lyle preferred that Carrie get both the exposure and the income from the work. He wrote to Carrie, "I wanted yours to get first shot, of course." She was paid a small sum for the article, a fact Lyle later lamented, saying he was "sorry that the combination of you-and-me is only worth thirty dollars."[26]

In the article Carrie described Saxon as one of those "few persons who step into legend while yet alive."[27] She wrote of Lyle having been one of the proponents of the resurrection of the Vieux Carré and how he was instrumental in bringing artists and writers back into the old homes of the French Quarter. She told of how he had bought a house there and furnished it with authentic antiques: "That was when crazy little antique shops in the Vieux Carré were packed full of real antiques, some of them priceless treasures now." Because of Lyle's example, she explained, others began moving into the Quarter as well.

When the article appeared in January 1931, a frustrated Carrie was dissatisfied with it, telling Cammie that "they messed up the Lyle article—whether accidental or not I don't know—but it spoiled it for me. I was glad you and Lyle had seen the original story."[28] The magazine's editorial decisions rankled her, but there was little she could do about it.

Lyle was back and forth between Melrose and Baton Rouge during the first half of 1930 but returned to the cabin in July for about ten days to do some work on the Lafitte book. "I'm in the worst fix that I've ever been as to the manuscript . . . my book is almost entirely unwritten and Century already wiring me to hurry with the manuscript. Now they threaten that they can't bring it out this fall if I'm later than August 15," he told Carrie.[29] He submitted the manuscript in September and a couple of weeks later wrote despondently: "I sit here waiting for a telegram from Century telling me that the proof is ready for reading. My mental condition is self-explanatory—for here I sit, sweltering in Baton Rouge—when I could be flitting north this minute, if I wanted to. Truth is, I don't want to go at all. Just want to die instead. I'm a wreck. Nothing, not even *Father Mississippi* was done under such pressure."[30]

Century had been "very patient" with him and allowed him to submit the Lafitte manuscript "in piece-meal, a chapter at a time," he said, and they were thrilled beyond measure when it was finally finished, as Lyle certainly must have been. *Lafitte the Pirate* was published in November 1930, received good reviews, and in 1938 it was turned into a film entitled *The Buccaneer*, directed by Cecil B. DeMille. But what Lyle still wanted most was to write his novel; the pressure on him—from both his publisher and himself—to get it done was overwhelming. He expressed these frustrations in a letter to Carrie in September 1930: "I wonder if I'll ever write a novel . . . ? Did I ever begin one? What was it about? Why? That's the way I feel. I need a drink. I need a lot of drinks. I need a crack in the neck, or to fall in love, or to break a leg or to eat ground glass."[31]

During the 1930s many southern states saw a rise in malarial infections. The neglect of drainage projects and loss of personal income during the Depression were at least two contributing factors.[32] The insecticide DDT was not used as a preventative until the mid-1940s. Malaria had always been a problem familiar to anyone who lived in swampy Louisiana or along its rivers and bayous. Cammie's family was no exception. Both Cammie and her mother had been afflicted multiple times with the disease. Sometimes Cammie was bedridden for days with it and would take quinine treatment, as was customary at the time, or even go to Hot Springs, Arkansas, for the rejuvenating baths.

In early August 1930, family friend Mitt Dunn came down with malaria. Since the death of his father, Dr. Milton Dunn, in 1924, Mitt had lived in the Melrose community and remained very close to the Henrys. When he became gravely ill, Cammie tried to oversee his care, but in one letter, she notified Carrie that "Mitt Dunn got so much worse I took him to Dr. Ragan yesterday."[33] The doctor began treatment, and a week later Cammie wrote, "Mitt dropped back— thought I'd have to take him to Shreveport today—called Dr. Ragan—says try treatment little longer—Mitt looks dreadful—and losing weight daily."[34] Another local physician, Dr. Keator, was called in and confirmed the diagnosis. Mitt's sister, Maude, was notified, and Cammie persuaded the family to take him to Shreveport for treatment. But it was too late. He died on September 4 at the age of thirty-seven.

As he would in the case of Fred Wilson, Lyle took the news hard: "You know, Carrie, I can't bring myself to write, even to you, about Mitt Dunn's death. I mean, I really can't. It was the most unnecessary, and heartbreaking thing I ever heard of. I had a note from Aunt Cammie, on the back of a forwarded letter, telling me that he died in a Shreveport hospital, on my birthday."[35] Cammie stayed in Shreveport long enough to help make funeral arrangements. Then she and Cammie Jr. went on to Murfreesboro to see Bobbie, after which she left Cammie Jr. at Winston-Salem Academy in North Carolina.

Cammie Jr. became desperately homesick at school that fall. She missed her mother, missed Melrose, and missed "Sam's morning coffee."[36] She and her mother exchanged letters almost daily, always with Cammie encouraging her daughter to concentrate on her studies and offering love and support. Letters from Leudivine to her granddaughter expressed the same sentiment. Carrie and Lyle often wrote to Little Cammie, and Lyle even sent her some cash for her fifteenth birthday in August. In November, Carrie told her, "You don't know how proud I am of you . . . I had no idea you would stick it out!"[37]

With the country firmly in the clutches of the Great Depression, Cammie struggled to pay the tuition for little Cammie's school. In 1930 the school account had a $385 balance. An itemized statement in Scrapbook 203 indicates that this amount included $345 in a general account, $25 for books, $10 for her "athletic costume," and $5 for personal expenses.[38] Cammie made money for renovation projects around the plantation by selling plants from her garden, rugs she had made, and other weaving projects. She boasted to Carrie that she sold "two dozen pink canna roots for $2."[39] Keeping up her customary unflappable outlook, she noted, "Life is worth living despite its heartaches."[40]

Still, stress from plantation finances, constant visitors, and family matters mounted, and that October, Cammie Jr. offered some uncharacteristically mature advice, encouraging her mother to go to Hot Springs and "take the baths." By the end of November, Cammie was in Hot Springs and lamented to Carrie: "Still have hives—Dr. has never located trouble—I'd have gone home with J.H. this past Monday—but he begged and insisted I stay a while longer—it's no good—of course the baths tone up any one—but I came to eliminate hives and the baths positively don't do it—well—I peg along a while longer—I bet if Dr. Ragan was well he'd cure me in a jiffy."[41] Almost a week later she wrote to Carrie again: "Well—I'm near the end—take my last bath Sun. Dec. 7—the hives are back in full force—body covered—but I do have a lotion that soothes the irritation—no signs of what I eat that causes this—Dr. says positively an intestinal disturbance—I've given it all a good honest try out—and the baths have helped my general health—so I'll try not have any regret—50 years hence—it will make no earthly difference."[42] At least while she was stuck in Hot Springs, she could occupy herself with reading Lyle's just-released *Lafitte the Pirate*. She declared, "I am charmed with *Lafitte*—it's good." She bided her time in Hot Springs, finished her baths, and returned to Melrose.

In the end boarding school was not the place for young Cammie. She left Winston-Salem in 1932, briefly attended Louisiana State University, and eventually earned her degree from Belhaven College, a small Presbyterian college for women in Jackson, Mississippi. A vivacious young woman and the only sister among seven brothers, Cammie Jr. quite possibly was indulged and spoiled. The age difference between the eldest son, Stephen, and Cammie Jr. was twenty years, and though Cammie Henry wanted her children to be affectionate and encouraged the boys in her letters to be kind to their sister, Cammie Jr. was naturally most attached to her brothers Dan and Payne, who were closest to her in age. Letters from family friend Robina Denholm to Carrie indicate that Cammie was unable to take a firm hand or a strong line with her daughter, noting, "Sister causes Miss Cammie much uneasiness—more than she admits I am sure."[43] In an undated letter from the early 1940s, Robina expressed her frustration not only with young Cammie's high spirits but also with the girl's mother: "It is hind sight but Miss Cammie is really to blame that Sister was never controlled for one instant in her whole life."[44]

After the Christmas 1930 holiday, Cammie got to work "like a house a fire" to get a four-poster bed "all rigged up with curtains" for Lyle's cabin.[45] Because Lyle was such a tall man, Cammie had the bed made seven feet in length, with "wood from the sides of an old cistern, the mattress made from wool grown on the plantation."[46] Leudivine was also involved in the project and told Carrie: "I'm saving feathers for a pillow for him to kneel on when he says his prayers. One of those Catholic affairs that looks so suitable at the foot of the bed."

In addition to her furniture projects, Cammie took to heart the advice of her late friend Henry Chambers and intensified her efforts to obtain volumes

for her library. She corresponded frequently with Otto Claitor, proprietor of a
bookshop in Baton Rouge, about books in his catalog pertaining to Louisiana.
In January 1931 Claitor wrote to Cammie that he would send her their "lists as
issued and keep you in mind for any Louisiana material of especial interest that
come to hand."[47] Over the next few months, she ordered a copy of Lyle's *Old Louisiana* from Claitor—she often presented Lyle's books to friends—as well as Kate
Chopin's novel *The Awakening*. In one of their previous discussions, Claitor had
recommended John Townsend Trowbridge's study *The South*, published in 1866,
but after closer examination, found it unsuitable: "I took it home and read it myself (as much of it as I could struggle through). Don't think you'd like it any more
than I did as it was evidently written for the especial benefit of carpetbaggers and
their northern radical friends of that period."[48] Cammie and Otto Claitor continued their correspondence for many years as she expanded her book collection.

In March, Ada wrote to Cammie about some books Cammie had apparently
requested she return. Ada insisted she never had the books and was troubled
that Cammie thought she did. "The only books I have of yours . . . if at any
time I borrowed the ones you wrote of (and I'm sure I did not, or I'd remember
it,—I've never read either), I returned them some time ago with the others." She
explained that she always had a "special shelf" in her house "exclusively for Melrose books." She went on to say, "Somehow I have felt that my last visit to Melrose (that hot, insane night last summer), was such a mistake."[49] It is not clear
to what she refers, but she subsequently made what Cammie called a "transient
visit" at the end of May 1931, spending one evening at Melrose.[50] Then a scheduled visit in June 1931 fell through. "Of course Ada did not come," Cammie wrote
to Carrie. "That calls for more talk—me & you."[51]

Later in the spring, Irene Wagner, a former English teacher at Louisiana State
Normal College in Natchitoches, visited Melrose. She was interested in Cammie's library, specifically her large collection of Louisiana books. Wagner had
been to Shreveport to examine the collection of historian J. Fair Hardin, and it
was through Hardin that Wagner gained access to Cammie Henry. Unlike Hardin, however, Cammie did not have a printed, bound catalog of her volumes;
in fact, her books were scattered all over the plantation in various guestrooms
and cabins. Wagner offered to compile an annotated bibliography of Cammie's
library that could be bound and serve as a permanent record of her books.

Because Cammie had so many books related to Louisiana, either by author
or subject, they decided to start there. Cammie gathered up all of the books

from around the plantation, placed them in boxes, and delivered them to one of the cabins, where Wagner was stationed with her typewriter. The books that were in the library downstairs remained on the shelves, and once Wagner had documented the stray books, she moved into the library to work. The entire process took about two weeks. Then Cammie's friend Robina Denholm, who worked as a stenographer in Shreveport, typed up the list and had it bound. Interviewed years later, after Cammie's death, Wagner said that when Lyle saw the volume, he remarked, "Well now, Cammie, at last we know what we've got!"[52] Cammie chuckled, and one might assume that she saw this catalog as bringing some gravitas to her library and making it at least equal to that of Hardin.

L yle returned to the plantation in March to take up work once again on his novel. At this time his publisher was still Harrison Smith, but that relationship had become strained because of Saxon's continued delay in finishing the manuscript. It is probably about this time that the oft-repeated legend about Cammie locking Lyle in the cabin until he finished his book originated. In 1931 reporter Alfonso Lahcar published the article "In the Land of Romance" in the *Shreveport Times,* in which he recounted his automobile trip through Cloutierville and then to Melrose. He wrote in romantic prose about the winding road hugging the river, the curves that "keep you guessing" about what would be around each corner, and finally the lush gardens of Melrose, where flowers are "growing everywhere, jonquils, flowering quince, yaupon trees, holly trees, wild ferns . . . japonicas, magnolia trees, huge oak trees . . . the beauty of all this just can't be written." Regarding Lyle, he noted that Cammie was "keeping him locked away from the public right now while he writes another one of those fascinating books."[53] Certainly, Lyle was not literally locked in the cabin, but it seems that some took it that way, and rumors began to circulate that Cammie was turning away his visitors.

It is true that Cammie attempted to keep Lyle undisturbed when he was working. In April 1934, three years after Lahcar's article, Lyle was still writing the novel. Cammie wrote of a "limousine full of people at 7 a.m. making for Lyle when servants told me and I turned them back—they furious."[54] The tourists apparently offered one of the hands money to take them to Lyle's cabin, but, she noted, "he had too much sense to do it." The next month two of Lyle's friends arrived one evening and went straight to the cabin. Cammie wrote: "I broke it up; forbid it—guests come to the boy's room. Lyle must not be disturbed."[55] Between the 1931 Lahcar article and the 1934 incidents, it is not hard to believe that Cammie probably did chase visitors away who might be intent on speaking to Lyle. She told Carrie in March 1931 that although she had never seen him "in better form—he has not plunged into the novel,"[56] revealing her worry about his lack of progress.

By September 1931 Lyle had finished his novel, but he was not pleased with the result and decided not to send it to his publisher.[57] His struggle with the manuscript caused the eventual dissolution of his publishing contract with Harrison Smith, and by 1932 he had contracted with H. C. Kinsey and Company.[58] His friend Roark Bradford, meanwhile, was enjoying much success with his latest novel, *John Henry*, which was serialized in *Cosmopolitan* and selected as a Literary Guild book for September 1931; on top of that, Bradford was receiving royalties from the Broadway play *Green Pastures*, adapted from his 1928 story collection. Fretting over financial woes, Lyle told Carrie: "Oh Lord, how difficult it is to keep on liking our friends in their prosperity . . . There's one sure thing, Carrie, you'll never have me put *that* strain on our friendship."[59]

Chance Harvey makes the point in her Saxon biography that Lyle spent a great deal of time procrastinating and occupying himself with pointless diversions, like training the Melrose chickens to land on his arms.[60] James Thomas says "his procrastination is undeniable," primarily because he was so concerned about getting the words "just right."[61] The plantation was both a retreat and a distraction for Lyle; while there, he could continue to present himself in the manner of a distinguished southern gentleman, receiving visitors and wiling away hours engaged in leisurely pursuits such as sitting on his gallery reading books or planning a white garden to plant behind his cabin. At Melrose he was seemingly a man without concerns. Certainly, he spent a lot of time corresponding. There are reams of typed letters from Lyle to Carrie Dormon from this period, for example. And he did spend time entertaining friends who came to Melrose and were deemed important enough by Cammie to let through the gates to see him, such as Sherwood Anderson, Edward Suydam, who illustrated three of Lyle's books, and Roark Bradford. In this way Lyle and Cammie both profited from their friendship: Lyle had a country home that suited his self-styled image as a sort of planter or aristocrat, where he could escape the social demands of the city; and Cammie had the company of Lyle and his entertaining and literary friends to amuse her and elevate her own cultural relevance through association.

I n the midst of much activity at Melrose in the spring of 1931, Cammie was laid up temporarily, as Lyle wrote to Cammie Jr.: "Your mother, poor thing, has sprained her ankle—not very bad, thank the Lord, but enough to keep her lying down, or sitting with her foot in a chair, and she hates the inactivity. But there she must sit for another week at least. She says that it doesn't hurt much, but I know that it must pain her a good deal or she would be up and running around. As it is, she sits in her room and gives orders to Bud and Sammie Peace and War Baby and Rosa the cook and to Mary and Allie . . . I go over and drink coffee with her every afternoon, and we have long talks."[62]

It is interesting to imagine what these long talks Cammie and Lyle may have had would consist of. Certainly, they had hundreds of intimate conversations through the years, and clearly they were two souls who were destined to be good friends. No doubt they talked about books, current events, and shared friends, but it is also likely they discussed reminiscences from the past and dreams of the future. Lyle's grandmother, Elizabeth Lyle Saxon (1832–1915), was a woman Cammie would have loved to know. A writer and early suffragist, she published her Civil War memoirs, *A Southern Woman's Wartime Reminiscences*, in 1905. Cammie undoubtedly enjoyed hearing about Lyle's family, and he about hers. Whatever their private conversations consisted of, the two surely always had in common a love for the South and its history, especially plantation life.

Easter, which always brought many guests to the plantation, that year saw some extra drama, which Lyle conveyed in his letter to Cammie Jr.: "We had a very exciting Easter. A lot of drunken negroes congregated somewhere down the road, late at night, and they fought so bad that someone came after J.H. Mr. Carter and Ernest, the overseer, went with him. The negro had a pistol and was threatening to shoot everybody in sight, white and black; so Mr. Carter took a pot-shot at him and shot him in the legs with birdshot. That was all there was to it, except that they had to take the shot negro to the hospital in Natchitoches. So nobody got much sleep—that is nobody but those who knew nothing about it—and I was one of them, worse luck."[63]

The number of visitors to Melrose in the 1930s increased as Cammie's library grew in reputation. The more articles there were in various newspapers or magazines, the more tourists appeared, and the more scholars or researchers wrote to her. This was sometimes a nice distraction, and Cammie did enjoy her correspondence with researchers such as Dr. Rudolph Matas, a prominent surgeon, and with other scholars, but sometimes she was impatient with the multitude of requests for information. She often remarked that if she went to all the conventions she was invited to or addressed all of the research inquiries sent her way, she would never accomplish anything.

One writer who came to Melrose was Ruth Cross, a native Texan who had spent a good part of her childhood in Winnfield, Louisiana. She needed to do historical research and gather material for a novel she planned to write about Natchitoches founder Louis Antoine Juchereau de St. Denis and the earliest days of Louisiana. Cross met with historian J. Fair Hardin and with Cammie, availing herself of both of their libraries. Her October 1931 visit to Melrose was reported in a *Shreveport Journal* article, which is preserved in one of Cammie's scrapbooks along with a handwritten note that says: "It doesn't take three geniuses to recognize St. Denis—and this is one episode I could blame on no one but myself that I didn't do it. I am only glad that the lady is a Texan, not a Yankee as I feared—and she has Spanish. She is undoubtedly well equipped? I shall be interested in the outcome."[64] It is not clear if Cammie herself wrote this comment or if someone else clipped the article and sent it to her along with the note. Cross spent two days at Melrose with Cammie's scrapbooks, and Cammie found her to be "very pleasant." Her novel *Soldier of Good Fortune*, her fourth book, was eventually published in 1936.

Cammie lent her pen as well as her library to support and encourage writers to tell about the South and its history, as in the case of John Uhler. In October 1931 Uhler, an English professor at Louisiana State University, published a controversial novel entitled *Cane Juice*. The protagonist is Bernard Couvillon, "the Cajun gorilla," who was born and raised on Bayou Lafourche. Bernard's father runs a sugar plantation, and sugar is not doing very well; pests like the borer and the mosaic are taking their toll, and the cane is not producing much juice. As it turns out, Bernard is a pretty intelligent fellow, and he is awarded a scholarship by the local police jury to attend the new university, LSU, where he hopes to attend the sugar school and learn enough to save the sugar industry in Louisiana. Bernard's father wants his son to stay and work in the mill with him, but Bernard packs his two-dollar suitcase and walks down the levee to LSU, where he is met with the ritualistic hazing by upperclassmen. The novel follows Bernard through his determined efforts to learn about sugar, his resistance to and later acceptance of joining the football team, and his relationships with the various characters he meets at the school.

The powers in control at LSU were strongly opposed to Uhler's book and promptly fired him. They found the mere suggestion of hazing offensive, and the representation of wild parties at abandoned plantations was considered scandalous. Even some in the Catholic Church were offended at the depiction of young people "breaking the Eighteenth amendment" and drinking alcohol.[65] Uhler enlisted the aid of the American Civil Liberties Union and within the year was reinstated on the faculty.

Cammie closely followed this story, as evidenced by her scrapbook clippings of each development, and she wrote a letter of support to Uhler. In response, he thanked her for the encouragement and promised to send her an autographed copy of the book, which was sold out all over the state.[66] Uhler had visited Melrose, but there is no evidence that he stayed or worked there in one of the cabins. One would expect his novel to appeal to Cammie as someone who grew up on Bayou Lafourche on a sugar plantation and also as the mother of sons who had attended LSU. Regardless of how she may have felt about the content, she certainly supported Uhler's right to publish it.

In May 1931 Lyle traveled to New Orleans to greet photographer Doris Ulmann, whom he had met in New York a few years earlier at a party with South Carolina author Julia Peterkin. Ulmann was a small, fragile woman, with short, dark hair and elegant manners. Lyle took Doris to photograph the nuns of the Ursuline Convent of the Sacred Heart, the Sisters of the Holy Family, and various sites in New Orleans before her chauffeur, George Uebler, drove them to Melrose. Ulmann's photographic expertise was in painterly portraits depicting generations of people she believed time would leave behind, such as Appalachian wood-carvers, Mennonite community members, and African American cotton pickers in Louisiana. She photographed boys packing cotton into sacks; old, wrinkled faces; close-ups of hands at work; and average people tending to their daily duties. She took one picture after another of the Melrose employees and the mulattoes of Isle Brevelle.

In a letter to Carrie, Lyle wrote: "We drove up to Melrose where Doris is taking pictures like mad. Or rather with a methodical saneness and intense activity, an average of about four dozen a day. She doesn't photograph places, ever, but takes only portraits of Sammie Peace with a cross-cut saw, etc. . . . She doesn't even glance toward acer rubrum or salix baby onions, except with a thought as to a suitable background (out of focus) so I suppose she wouldn't interest you much, but her figure studies are really superbly good."[67]

Ulmann's trip to Melrose would open a correspondence with Cammie, who made a striking impression on the photographer. Upon her return to New Orleans, Ulmann immediately wrote to thank Cammie for the hospitality, saying the time at Melrose seemed "like a beautiful dream that took place in a delightful country."[68] Enchanted, Ulmann went on to say that her visit "has been one of the precious events of my life that I have had the privilege of knowing you and I am deeply grateful to Lyle for having brought me to you. I have not known you very long but you must believe me when I tell you that I have been missing you intensely." The high regard was reciprocal; Cammie found Ulmann to be an exceptional woman and enthused about her to Carrie: "Doris Ulmann is splendid—perfect—oh I wish you could know her & see her work. Works like a beaver—and with all those millions—simple & gracious—speaks several languages fluently. She is tip-top—I would not take any thing for know her [sic]— Her pictures! Well, you never saw the like & she does it all herself—takes, develops, and prints. Exquisite work."[69]

It is not surprising that Ulmann and Cammie formed a friendship, as they were engaged in a common purpose, although through different mediums. Ulmann's photography was focused on preservation of lost cultures and types, not unlike Cammie's projects at Melrose. Ulmann and Lyle also continued a friendship for the duration of her life. He sent her small birthday gifts, on her last one a trinket box made from a book. A woman of secure means, she would on occasion send Lyle a check to ease his financial troubles. In June 1934 she wrote to him asking if he was at least "tolerably happy," adding, "Lyle, my dear, I also want to know whether you have been starving—you know very well that I object to this."[70]

Cammie and Lyle were shocked in September 1934 to learn of Ulmann's sudden death at age fifty-two. She had been in declining heath in her last years and had battled orthopedic and respiratory problems her entire life. Eventually, she succumbed to kidney problems and blood poisoning, according to her biographer Philip Walker Jacobs. Ulmann had continued to work as much as she could and refused to slow down, relying more and more on her companion and assistant, John Jacob Niles. Lyle was at Melrose when he received a tortured letter in broken English from George Uebler full of mysterious insinuations about her death. Lyle was terribly upset by this letter and wrote back immediately to get more information. He also contacted Julia Peterkin, who wrote that the whole affair "is simply too painful to think about" and that she had taken refuge at Lang Syne, her plantation home in South Carolina.[71] Cammie and Lyle stepped back from the mystery, leaving it to parties more directly involved, but the loss of Doris Ulmann was one that they both felt strongly.

Nearly two years after the stock market crash, in August 1931, Cammie wrote to Carrie, declaring: "The bottom done dropped clean out of cotton! J.H. has quit eating. What good will that do? I wonder where it will end."[72] Cammie and Carrie were both enterprising and clever and continued to find ways to pick up a few pennies; for example, they would sell copies of Carrie's blueprints to the Melrose tourists. Cammie told Carrie, "I was delighted to sell the blue prints—I've got to make something—utilize these people who come to my door—If I can make enough to buy a loom I might sell my product."[73] It also appears that they gave thought to selling chairs or stools made by a local craftsman with cane-bottom seats to be woven by Cammie. In an August letter, Cammie wrote: "If I lose everything else—I've got you! And after all, outside of companionship, so little matters . . . I wonder if we could sell home made chairs, seasoned wood, nice bottom—those tall back rockers if he made for $2.50? I wonder if we could not sell for $5.00?"[74] Cammie's revenue efforts were not always well received by her children. "I expect Dan will die if I begin to sell things," Cammie said. "He will just have to die—I can't sit here with folded hands—life is too short to be idle." Even Leudivine confessed to feeling the financial pinch in a letter to Carrie in October, saying, "We are scrambling for money too—some things just have to be met with money."[75]

Because of the tight financial situation, Cammie also had concerns about

Bobbie at Cora Bristol Nelson's school in Murfreesboro. Bristol Nelson and Cammie had become friends through the years, exchanging letters, and Bristol Nelson even made a social trip to Melrose in February 1931, spending five days at the plantation, after which she wrote Cammie that "years have been rolled away—burdens have dropped from my shoulders—cares that almost overwhelmed me have lifted."[76] She took an autographed copy of *Old Louisiana* home with her. The visit, she said, was "my idea of perfection here on earth."

The Bristol-Nelson School was struggling under the financial crisis and had to make exceptions and allowances to keep its students. Bristol Nelson agreed in September 1931 to keep Bobbie, by then almost twenty years old, enrolled for half of what Cammie had been paying. She told Cammie, "You have stood by me too long," and made arrangements for the Henrys to pay thirty-five dollars per month, deferring the balance of the monthly payment and letting it accrue for two years without interest.[77] This was a tremendous load off Cammie's mind, not just regarding her pocketbook but also in terms of Bobbie's stability and well-being. He was well adjusted and happy at the school, and neither Cammie nor Bristol Nelson thought it was in his best interest to be moved. Cora explained: "Bobbie gets so nervous with many children—loves to be alone with his little toys . . . he dislikes roughness . . . But no arguments—Bobbie is mine and must stay right on here—so don't worry."[78]

Stephen visited Bobbie regularly, and in July 1931 Bristol Nelson wrote to Cammie to tell her how much Bobbie loved his visits. Cammie noted in the margin of this letter: "Bobbie is well and happy. All we can expect."[79] Ten years later, in 1941, as Cammie grew older and in anticipation of the future without her, J.H. took legal control over Bobbie's affairs through the court, receiving a judgment declaring Bobbie "incapable of taking care of his person or administering his affairs."[80] That Bobbie be well taken care of and kept happy was of primary concern to the Henry family, and this judgment likely would have given Cammie some peace of mind to know that J.H. would continue to take care of Bobbie in her stead.

In January 1932 the woman who brought Cammie and Lyle together in her literary salon years earlier died at the age of seventy-nine—author and "la grande dame" of New Orleans, Grace King. Lyle served as a pallbearer and spoke at the funeral: "I cannot speak of a memorial for Mrs. King, for it seems to me that in her work she herself has created a memorial that surpasses anything that we might build in her honor."[81] King's passing caused a profound bereavement for Lyle, who admired the woman and her work very much. Cammie wrote to Carrie: "Mate—Well—Grace King died yesterday. What a loss! Of course Lyle took active part . . . we did not go to N.O. . . . This death will upset him terribly."[82] It was only the beginning of a rough year for him.

In the course of the following six months, Lyle became exhausted battling a series of serious health issues, including syphilis, which forced him to undergo painful and intrusive treatments with his physician in New Orleans.[83] His biographer Chance Harvey writes that Lyle "had become less discriminating, even careless, in his choice of sexual partners," and biographer James Thomas indicates he "had intimate relationships with both men and women."[84] In a July letter to Carrie, Lyle wrote: "Hey God, but I've had me a time this year. Pains and miseries; and treatments from a doctor which reduce me to such depths of humiliation that I blush to think of them. I'm damned sick of it all—six months, and still I fall to pieces and swell up and hurt . . . I've been in constant pain since the thirteenth of January."[85]

As miserable as he was and as poorly as he felt, he still offered encouragement to Carrie, who was working to get her book *Wildflowers of Louisiana* published. In the same letter, he told her: "I'm most anxious to read your book—and if I can do anything with reference to publishers, you know I'll be delighted. I know some of those critters fairly well, and a note from me might insure a friendly reading."

In September, Lyle fell victim to a terrifying robbery. He was in New Orleans at his Royal Street residence when two men forced their way inside and demanded two hundred dollars. When he told them he did not have that much cash, they forced him to write a check for one hundred dollars and then to phone the clerk at a nearby hotel with approval to cash it. One of the robbers held Lyle hostage while the other went to cash his check. Cammie learned of the robbery through the newspaper, and in a letter to Carrie on September 18, she said that one of the robbers had run errands for Lyle in the past.[86] Lyle described the event to Carrie in October: "I was cracked over the head and robbed . . . the nervous shock has done something extraordinary to me, coming on top . . . of all those long months of pain. I'm so perilously near some sort of breakdown that it isn't even funny. The trial of those gangsters comes early next week, and I must go through with that; it will be an ordeal."[87]

According to Lyle, the two had been "using every type of intimidation" to keep him from testifying, including calling him constantly on the telephone each night, "usually about three a.m.," threatening violence. "This has kept up for a month," he said. He could "no longer sleep" and said "the flapping of a window shade makes me shake for half an hour." The two men eventually pleaded guilty to larceny, but Lyle remained traumatized. He was looking forward to getting out of town and going to the quiet of Melrose when possible, and he repeatedly planned to go but seldom followed through. In early November he sent word that he needed to be picked up at the train in Derry, but when Cammie sent a driver for him, Lyle had not come after all. "According to orders," she wrote to Carrie, "sent to Derry for Lyle last night—no Lyle—no wire . . . I pray he is not sick."[88] She prepared the cabin for him and waited.

That November, Karle Wilson Baker, a poet of note from Nacogdoches, Texas, visited Melrose. Baker had published fiction and poetry in *Harper's,* the *Century,* and other contemporary literary magazines as well as two popular books of verse, *Blue Smoke* (1919) and *Burning Bush* (1922). Her 1931 poetry collection, *Dreamers on Horseback,* was nominated for the Pulitzer Prize. Baker had learned of Melrose in March through a feature story written by William Goolsby in the *Dallas Morning News.* Anxious to gain entry, she wrote to Goolsby, who then wrote to Cammie, asking her to issue an invitation. Baker brought her husband and her daughter, illustrator Charlotte Baker Montgomery, and made several more visits to Melrose over the next year. She met Lyle Saxon there and gave Cammie and him signed copies of her books.[89]

In the spring of 1933 the house was full of more guests, prompting Cammie to tell Carrie that she seemed "to do nothing yet use up every ounce of strength daily."[90] The gardening chores were endless, and Cammie also stayed busy directing operations in the kitchen, refinishing old furniture, making rugs and curtains for the house, and doing countless other chores. In March her guests included Josephine Grunewald, wife of Theodore Grunewald, who had owned the Roosevelt Hotel in New Orleans; Maude Dunn, sister of Dr. Milton Dunn, for a week;

and Shreveport friends Clift Byrd and Robina Denholm for a weekend. Cammie put them all to work on various tasks. "That's my long suit—giving jobs!" she wrote to Carrie.[91] Robina, for example, enjoyed refinishing furniture. Cammie was also expecting Lyle, and artist Alberta Kinsey was due to return in April.

While Ada seldom visited, Cammie's friends kept her apprised of Ada's activities. In May 1933 a devastating tornado hit Minden. Ada reportedly saw the storm approaching from the southwest and phoned her husband at work to alert him. John and son David took cover in a culvert near the mill and were thankfully uninjured.[92] Mary Belle McKellar wrote to Cammie: "Saw Ada at the Armory. The next day Mr. S[nell] said she was working at the Relief station. Said she had been there since 6 o'clock in the morning. Was glad to go, of course, that she could get material there, too. Perhaps this tragedy although it didn't touch her personally will stimulate her talent once more."[93] Ada had not published anything since 1928, a fact that generated much concern and curiosity among her friends, and apparently even a volunteer shift at the Armory could be seen as a source for new material for a story. More likely, Ada was simply fulfilling what she saw as an obligation to her devastated community with her volunteer work, as she had been involved in the Minden social clubs and with her church for years.

When Lyle at last returned to the cabin that spring, he stayed the majority of the summer and fall and resumed revisions on his novel. He told Carrie that he had "been reading the first hundred pages of my novel as it comes from the typist. It is better than I thought, but strange. Can you imagine my writing passionate love scenes on the River Bank?"[94] These first pages of the novel apparently satisfied him well enough to send them to Kinsey, his publisher, in October 1933, although he was nowhere close to finishing the book. He was still vulnerable to the distractions afforded him at Melrose, which are detailed in long, gossipy letters to Carrie about there being "too many outsiders" to visit properly. He complained about a woman who "slouches in chairs with her legs cocked up," casting him alluring glances, and he implored Carrie to "tear this up, for God's sake!"[95] There are more letters to Carrie Dormon from Lyle during this extended period of time at the Melrose cabin than almost any other time in their friendship. Filled with minutia and trivialities, they are a testament to his ongoing inability to focus on his work and his willingness to embrace almost any possible diversion from it.

Once Carrie's *Wildflowers of Louisiana* had finally been accepted for publication by Doubleday, Doran, and Company, Lyle spent the next few months consumed with helping her look over contracts and negotiate with the publisher. This was a kindness to Carrie, but it was also just one more thing to keep Lyle from completing his novel. When word first reached Melrose in August 1933 about Carrie's book, it came from Cammie Jr., who had descended on Briarwood with a group of her friends and learned the news from Carrie. Cammie Jr. promptly came rushing back to Melrose, shouting that Carrie had "a book accepted by Doubleday-Doran on wild flowers."[96] Unable to contain his enthusiasm, Lyle wrote to Carrie: "Lord, but I do hope that that part of their scrambled stories is true. And, if so, I'm even more glad than I would be if it were some similar success of my own." Cammie and Lyle waited to hear from Carrie herself, but the next morning a mutual friend called Cammie to tell her about a long story in the *Shreveport Times* that included news of the forthcoming book.

"Aunt Cammie tried her best" to find out more, Lyle wrote Carrie, "and came hot-footing it to the cabin . . . with this additional news. Naturally we are delighted, and toasted you in coffee as we sat surrounded by chickens in the swing on the cabin gallery." Cammie also wrote to Carrie: "Hurrah! If what *Shreveport Times* says is true—do tell us—if your book goes over we are delighted—do write us all about the contract."[97]

Contractual negotiations eventually stalled with Doubleday, Doran, and a benefactor stepped up for Carrie. Friend and philanthropist Edith Stern from New Orleans committed to underwrite the publication of the first run of *Wildflowers*. "Write them at once," Lyle advised Carrie, "giving her address and her offer, and tell them, too, how many signed subscriptions you have in hand." Lyle also advised Carrie as to the ownership of the prints in the book: "The pictures belong to you when Doubleday finishes making reproductions of them. [Edward Howard] Suydam always retains his sketches, and sells them afterwards, when he can."[98]

By the end of 1934, Carrie's first book was ready for market, and the Melrose group was excited but nervous. Carrie and her sister, Virginia, drove to Melrose to spend the night in December, bringing copies of the book with them in the trunk of their car. Cammie declared the book "lovely" and wrote: "She deserves success. The book has been underway seven years."[99] The group was anxious about the timing of the book's release, however. "We tremble," Cammie said in her diary, noting the number of books that must be sold. "It scares us—for what will sell in January?—nothing, everybody broke." Their fears were not necessary; the book sold quite respectably and was well received. Carrie and Virginia traveled the state—Shreveport, Alexandria, Baton Rouge, and New Orleans—selling copies wherever Carrie's Ford could take them.

In 1934 Cammie experimented with keeping a diary. She was an avid reader and collector of old diaries for her library, and in later years she and François Mignon would make it their work to transcribe and preserve diaries of the antebellum South. Perhaps this is what Cammie was trying to do with her diary, realizing that someday people would want to know what life on a plantation was like in 1934 rural Louisiana. It also could have been the fulfillment of a request by Lyle; just inside the bound volume, which is covered with a rough red-and-white handwoven material, Cammie wrote: "At your request here is a year at Melrose. Individually and privately yours. Lyle from Aunt Cammie, July 19, 1935."

Despite Cammie's inscription of *privately*, it is clear that this diary was intended for public consumption. Throughout the diary, she often assigned identification tags to people who should be obvious had the volume been intended strictly for Lyle's eyes; for example, in July she said, "I was sorry not to have Mamie Gray (Stephen's wife), Bubby their son, and Sister" for dinner. Any mention of Victoria Bebe Baptiste is always followed by "(cook)." On several occasions she even identifies her own grandchildren, such as in her Thanksgiving entry, when she says that it was "nice to have Joe, Eugenia, and precious little Pat (J. M Henry, Jr.)" in attendance for dinner.

In contrast, Lyle's journals are often small, slim volumes that he would be able to tuck into a pocket as he traveled or covered various stories for the newspaper. Lyle's inscriptions are quite personal and clearly not intended to be viewed by the public, as he occasionally shares his intimate perceptions of various people, which are not always kind or flattering. He was also forthcoming about his frustrations and feelings; whether he was elated or in the deep bowels of depression, it was recorded. Sometimes Lyle would go long stretches without an entry, while Cammie was diligent about recording something each day. Lyle's diaries span several years throughout his life, and Cammie only tried the experiment for one year. And finally, Cammie's diary was typed and then bound in her bindery. She even pasted in photographs to show various buildings and people of the plantation. Given all of this, one has to take the perspective that she may not have always been recording her personal thoughts and observations but that she intended the volume to be only a public record of the day-to-day routines at Melrose for one year.

It is clear that her diary was intended for posterity, whether "private" or not, and like Lyle's diaries, it remains an important record of life at Melrose as Cammie saw it. It begins rather formally with brief entries recording visitors and weather. The entry for January 13, for example, says: "Lyle went to New Orleans on business. Brother came to see Mother." The one for January 18 reads: "Cora Messiere spent three-fourths of a day. Afternoon, Mrs. Ada Jack Carver Snell, Rev. Sam Slack, Mrs. Stokes . . . Rain in P.M." Her letters to friends and family written during the same period are much more personal and informal. On March 19 Cammie simply recorded in the diary: "Three race horses arrived—Charles Mazurette bought them in New Orleans. So cold they had to have special protection. Pipes all drained, also cars. It promises to be cold." On the same day, Cammie's letter to her daughter is obviously more reflective of her natural voice: "The race-horses bought by Mr. Mazurette came in fine style—perfect house on wheels—made a stir on Cane River. I hope they prove successful."[100] She went on to say in the letter that Carrie had been at Melrose over the weekend with Lyle, Robina, and Clift Byrd, which "made a fine group." She wrote that J.H. was "crippled up" after getting pitched by his horse. "J.H. is reckless," she added. None of those personal details are in the diary. However, it does reveal a rare personal indulgence and a rare indulgence in modernity: in May 1934 Cammie treated herself to a "lovely new Singer sewing machine" to replace the model she had been using for twenty-five years. Sewing on the new machine "is a treat," she wrote.[101]

As the year went along, perhaps Cammie became more comfortable with her diary, or perhaps she let Lyle read it and he may have suggested she liven the thing up a bit. By the latter part of the year, the entries become much more conversational; for example, the November 26 entry reads:

> Cold. Dan went to Shreveport for the day (this was a holiday for him). Nice to have Joe and Eugenia and precious little Pat (J. M. Henry, Jr.). Lyle is always a pleasure, and adds 100% to the life at Melrose—be a tremendous

vacancy in my life if he dropped out. Let's never strike oil and Lyle never go to Hollywood. Both would be the ruination of Melrose. Rained a good deal last night, fair today. Henry Hertzog finished covering his house—storm last week lifted the tin. Began a gay rug à la Saxon, weaving it with plain weave and colored stripes. Rain all day, storm in P.M. Joe and I planned to go to dentist, but too much rain and wind. J.H. was run into by a truck at mouth of Montgomery lane, tore his car (was Joe's) all to pieces—a narrow escape for J.H. My, I am thankful he was not hurt. It is nice to have Joe and his family. Pat so happy here.

Cammie was never fully at ease recording her daily life. In the October 13 entry, for example, she wrote, "You people who are crazy about diaries, just try keeping one for even one year, then you will understand and not keep saying 'Why don't you tell more?'" Cammie was a woman of action, of accomplishing things, and while she enjoyed reading and preserving diaries she believed to be of historic relevance, keeping one of her own seemed to interfere with her actually living life. By the end of the year, she was no more pleased with her experiment, noting: "The end of this diary about the drabbest thing I ever saw. All of it not worth the energy Robina put on it . . . it's worthless and I'm tempted to destroy it."[102]

The diary records the cotton profits for that year: "J.H. and some Natchitoches men left at 5 A.M. for Baton Rouge to try and have tax on cotton reduced. Jady back at 7 P.M.—no success, tax $27.00 per bale; means about $6,000 for Melrose. Right now we are paying for all these unemployed who have no jobs."[103] Additionally, some of the Melrose restoration work is documented. The June 7 entry reads: "Henry Hertzog has begun on the 'shop.' Going to be lovely. Got Lyle to guide, spent two days cleaning it out. Was once a kitchen, then blacksmith shop, then lumber house. Building is old, put together with wooden pins. It has great possibilities. Henry Hertzog working but bent double with pain in his side."

Later, in August, Cammie and Robina Denholm took a quick trip to Navajo Plantation just above Natchitoches to look at an old house. Cammie was taken with a little one-room log house on the plantation that had hand-hewn timbers, and so in September she sent Lyle and Robina to look at it again. In October, Cammie and Lyle went back once more, and in her October 4 entry, Cammie detailed how she acquired the cabin for Melrose: "At Navajo house we met a one legged mulatto who asked to ride back to Natchitoches with us. In conversation he told us he was sure Mr. Herman Taylor (owner of Navajo Plantation) would be glad to sell the log house. We stopped and saw Mr. Taylor and bargained to get the old house for $25.00—cheap. Hope to move it next week, will make a lovely loom room."

Within the week she had dispatched Melrose hands Henry Hertzog, Puny Count, Haywood Collins, Jack Marcel, Manuel Hunter, and Nathaniel Brew "to the Navajo place to take down the log house." She "sent them off with a big truck and two Thermos jugs of coffee and a jug of water." The crew spent the entire day working in miserable rain to disassemble the cabin and bring every brick and log back to Melrose, including the chimney. Ironically, Navajo Plantation burned just two weeks later, a loss Cammie recorded: "Pity. What a fire it must have made—had marvelous timbers in it, said to have been easily 150 years old. Thus go the land marks."[104]

By the end of October, plans to restore the log house from Navajo were in full force; one afternoon Joe Ledet (the one-legged mulatto) came to dinner at Melrose and negotiated a deal to rebuild the chimney on the log house for Cammie for eight dollars. "After much discussion (and doubt on my part), I consented," Cammie said.[105] True to his word, Ledet showed up the next day to begin work on the chimney.

With Henry Hertzog overseeing the work, reconstruction began on October 18 and was going along just fine until the twenty-first, when Henry was arrested for drunk driving in Shreveport. He "tore up a banker's car, damage $150.00. Of course clapped in jail," Cammie wrote in her diary.[106] The incident slowed the work, but Cammie went ahead and got cypress posts for the gallery and green moss to be used in daubing the walls. On October 26 she recorded: "Jady agreed to have Lyle go to Shreveport to see about Henry Hertzog. It was the thing to do, so Lyle and Fuggerboo set off around one o'clock. Trip a perfect success." Lyle and Henry headed back to Melrose but not before a stop at the saloon at the bridge for a celebratory drink for all. There were a couple of drinking establishments on Cane River around Melrose: one was Frenchie's, which was later made famous in an iconic photograph by Farm Security Administration photographer Marion Post Wolcott; another was Bubba's (pronounced on Cane River as Buh*buhs*), which did not open until 1942. With the end of Prohibition, there was always a place to get a drink.[107]

J.H. loaned Henry the money to make the first fifty-dollar reparation payment to the banker whose car had been damaged, and with the incident now
behind them, construction on the log house began anew. Henry, as head carpen-
ter, directed J. C. Williams, Manuel Hunter, and War Baby onto the roof to begin
shingling, while Cammie and Lyle sat on the side as guides.[108] The stonework
was finished by November 8, and Cammie noted the "gallery pretty bad, but
eventually it can be remedied." A little concrete threshold connected the stone
gallery to the log house, and the 1934 date was written in the wet cement, where
it can still be seen today.

Cammie would spend a great deal of time in her new loom room, where
she placed several of the many looms and spinning wheels that she had collected. Her weaving hobby had blossomed; she had moved from weaving cov-
erlets and bedspreads to producing wool rugs and selling them to fund her restoration work. In June she sold three of her rugs for thirty-five dollars each, a
nice sum. She also used her work to furnish Melrose; she gave Lyle ten yards of
red upholstery to take to New Orleans to get a chair covered for his cabin. The
upholsterer, Lyle said, loved the fabric and offered him ten dollars a yard for it,
which he refused. "I would have sold," Cammie said. And with that project accomplished, she laced her loom to make curtains.

Cammie's interest in weaving was a serious one and attracted prestigious
names in the field to Melrose. Rudolph Fuchs and Kenneth Hunt, from Denton,
Texas, wired in mid-November 1939 to say they were headed over to stay at the
plantation. Fuchs was an associate professor of art and taught weaving at North
Texas State College. Hunt was also on the art faculty there. Fuchs studied weaving with the Weavers and Spinners Guild in England and eventually exhibited
his work all over the country. The two met Cammie through a common friend,
Gustine Weaver, who lived in McKinney, Texas, receiving their first invitation to
Melrose over their Easter vacation in 1937.

Their first morning at Melrose was gloomy, with a cold drizzle falling, so
rather than tour the gardens, Cammie instead took her guests to the loom room,
where they fell into a discussion on the durability of ramie, which Cammie was
experimenting with in her garden. While Cammie grew ramie plants, there is no
evidence that she ever developed them into material for weaving. Fuchs noted
that Cammie had tried her hand at raising the yellow-brown nankeen cotton,
which keeps its yellowish hue after weaving. According to Fuchs, this experi-

mont "did not meet with the whole hearted approval of those in charge of the plantation, because the yellow cotton will 'taint' the white cotton easily if they are planted too close together. Also, the plantation gin had to be cleaned out thoroughly after the yellow had been put through" before the gin could resume its regular run. Fuchs wrote that Cammie's method of learning about weaving was "disguising her knowledge of the subject" so that others would tell her what they knew. They would also "answer her many questions about loom construction, draft writing, new materials, and many other aspects of weaving."[109] Cammie produced many household items for Melrose, including bedspreads, upholstery, draperies, placemats, and rugs.

With renovation of the loom room complete, Cammie undertook a project to make Lyle's cabin as inviting as possible. She was elated that he was spending so much time there that year, and she felt that he was finally making progress on his novel. J.H. helped her get some lumber to redo the flooring in Lyle's cabin: "Hauling flooring and ceiling for Lyle's cabin. J.H. got it cheap, odd length, (5 feet) but No. 1 lumber—going to be splendid. If Lyle just stays here I'm fixed."[110] Cammie set Henry Hertzog to work on Yucca in early December 1934. "At noon Henry Hertzog and Haywood [Collins] begin to over-haul Lyle's cabin, will begin on [Edward Howard] Suydam's room—ceiling first. Will be fine when done. Lyle is going to have to move temporarily into the shop or log cabin." And move he did. Cammie's entry the next day reports: "Lyle found out he had to literally move out during repairs to his cabin—went into log house with his things. His arrangement makes the place look like something else, and most inviting. All went well until about 7 p.m. when mud chimney began to burn at back and on outside—mud too thin. Lath that held mud cats began to burn. We wet and fixed it, have hose ready to turn on stream of water, so all looks safe."[111] The chimney gave them no more trouble.

Another of Cammie's projects was the famed portrait of Grandpère Augustin Metoyer. The almost life-sized 1836 portrait was in the possession of his granddaughter, Madame Aubert Roque. For years Cammie had tried to barter with her for the painting, promising to have it restored and the jagged gash in it repaired if she would let it hang at Melrose. The portrait was painted in 1836 by the artist Feuille, who lived in New Orleans with his brother, an engraver.[112] Madame Roque finally agreed in late 1934, and in 1935 Lyle turned the portrait over to New Orleans artist and art restorer Frank Schneider, who completed the restoration project by the end of the year for a fee of one hundred dollars.[113] In January 1936, fully restored, Grandpère was back at Melrose, where the painting hung in Lyle's cabin; a smaller version of the portrait was also created and given to Madame Roque.

Cammie Henry's penchant for collecting antiques was aided by family friend Charles Mazurette, who lived in New Orleans. In early May 1934 he came to Melrose and brought Lyle's friend Olive Lyons with him; it was her first visit to Melrose since 1925. Mazurette had recently purchased a farm on the Little River, about five or six miles to the back of Melrose, and had stayed as Cammie's guest at the plantation while he worked on building a house. In May he came to Melrose bearing gifts: some antique serving pieces and "a lovely red glass shade for over the dining table."[114]

Whenever Cammie was looking for a specific piece, such as a fender for a fireplace or a soup tureen, she would write to Mazurette in New Orleans and have him search the shops. If he found a piece he thought Cammie would like, he would sketch the object in a letter for her approval and offer his advice as to price and condition. With regard to a fender he had located, he noted it was in good condition but not a real bargain: "We might do better over a period of time watching for someone who is hard up, or maybe picking up one at an auction sale."[115] He also found a Sheffield soup tureen for thirty-five dollars that he was able to negotiate down to thirty dollars. Cammie apparently gave her approval because a week later he wrote, "You are now the proud owner of a genuine Sheffield tureen and a very pretty fender."[116] He promised to ship the fender to her because it was too large to fit in his car, but the tureen "is too precious to let out of my sight," so he brought it to her personally. Mazurette loved spending time at Melrose and wrote to Cammie, "Melrose means you, the glow of the log-fire, and all the things you have to tell me."

Even Cammie Jr. sent Mazurette on antique hunts. Shortly after the fender and tureen purchase, he was dispatched to find a Kappa Alpha fraternity pin but came up empty-handed. "It seems that KA's do not pawn their pins," he wrote. He was on the lookout for bathroom fixtures and a tub next and said, "The only difficulty might be in getting one large enough for that big fellow Lyle Saxon."[117]

Mazurette's friendship extended to opening his home to Cammie Jr. when she had to have dental surgery on a wisdom tooth in New Orleans in June 1934. It was a difficult and painful surgery, and she stayed with his family to recuperate. Mazurette sent frequent wires and letters to Cammie at Melrose to keep her apprised of her daughter's condition and reassure her that young Cammie was just fine. "You certainly should know that without further assurance, that we are happy to have Sister here with us," he said. "She seems perfectly satisfied with her adopted parents and we all get along fine together."[118]

During the May visit at Melrose, Cammie and Olive spent time refinishing furniture, sightseeing, walking the gardens, and even tending to mundane tasks such as taking Cammie to the dentist. Carrie Dormon came down one afternoon, and the next day Olive's husband, Clifford, joined the party. In her diary entry for May 12, Cammie wrote: "Hilarious crowd at supper—love to see them spar at each other—J.H., Dan, Payne, Lyle, Clifford, Olive, Charles Mazurette, Frances McBride. Of course Victoria (cook) frustrated." Lamenting poor Victoria's kitchen skills, especially when serving company, Cammie noted that she "is better than an efficient stick."[119] The next day brought Alberta Kinsey to Melrose and saw Clifford and Olive head back to New Orleans. The revolving door of visitors continued throughout that spring and summer.

Many of the visitors came to see not just Cammie but also Leudivine. Mrs. Garrett had a warm friendship with both Lyle and Carrie, frequently exchanging letters with Carrie, who often sent a small plant or flowers for her room. Cammie's 1934 diary documents the final decline of her mother, then ninety-four years old. Throughout the year, she wrote, "Mother not so well" or, conversely, "in fine form." Cammie had been worrying about her mother's health for at least a decade; as far back as March 1926, she wrote to Carrie that "Mother was very weak—far away look in her eyes" and had come "mighty near the brink" of death. She was, according to Cammie, "failing fast—such a tragedy for me."[120] But Leudivine would live nearly ten years after that particular illness.

Lyle was very fond of Leudivine and saw in her a certain mischievous spirit, but by 1933 he, too, was worried about her health. In a letter to Carrie in November, he said: "Mrs. Garrett is slowly dying—or so it seems to me. Each day a little further off, each spell of weakness leaves her so further from life. And yet, it is all so natural, so in accordance with nature . . . that I cannot feel any acute pain, only regret to see her go."[121]

In May 1934 Cammie took a day trip to Alexandria, where she and a friend spent an afternoon touring central Louisiana to follow up on leads about looms and local practitioners of weaving. They went to Echo, Marksville ("had dinner there"), Mansura, Poland, Evergreen, and Hersmer, then returned to Alexandria. She was back at Melrose by five and in her diary noted that "mother had done fine."[122] Lyle wrote to Carrie that it had been "a peaceful day," which he had spent writing: "Aunt Cammie has gone off with Fuggerboo to Alexandria, and she and Mrs. Evans are going to see some weaving or something. The minute Aunt Cammie was gone, Mrs. Garrett arose and washed her hair . . . and has been parading around all day with the greatest gusto. She seems perfectly well today, and I only hope to God that she doesn't decide to go downstairs in Aunt Cammie's absence."[123]

By fall, however, Leudivine's health truly was failing. On October 5 Cammie was alarmed enough to send a wire to her brother, Stephen, in New Roads. He seemingly dismissed Cammie's concerns, saying he "had to work and was awful tired." Displeased, Cammie wrote in her diary, "Sorry I sent the wire but mother was so low I felt it my duty to send it—won't do it again."[124] Stephen had been to visit Melrose with his wife, Aimee, and their children in the spring, and he brought along a little fox terrier that added to the Melrose menagerie of dogs, cats, chickens, peacocks, and the customary farm animals. It is quite possible that he was unable to simply drop his responsibilities as a parish engineer in New Roads and run up to Melrose every time Cammie thought Leudivine might be on her last breath; Stephen had almost certainly received fearful alarms through the years. In the absence of her brother's assistance, Cammie sent for the doctor, hired a nurse to help her at night, and soldiered on. Despite his brusque response to Cammie's wire, Stephen did come up to Melrose to check on his mother and sister. Cammie noted in her diary that her brother looked "thin and not very well," which could also explain his reluctance to travel to Melrose. He spent only the night before returning home, leaving "immediately after breakfast" the next day.

Leudivine rallied and was up and down over the next couple of months. The second week of November, she was quite ill again, and Cammie was dissatisfied with the treatment prescribed by Dr. Scruggs. She called in Dr. Bath, who said,

"By all means do not let her suffer," and administered an opiate, which, according to Cammie, "was her salvation." She noted in her diary that her mother's regular doctor, Dr. Scruggs, would "never do this": "Thanks to Dr. Bath, Mother had a marvelous night after a terrific spell of suffering, an eighth of a grain of morphine saved her life. Dr. Scruggs could not see it. She had to rest or give out—all today she has rested; actually sat in the sun on front gallery, has not needed ammonia."[125]

Again, Leudivine improved, received callers, and dismissed her night nurse. By mid-December, though, she had to have morphine again. Cammie's December 17 diary entry states that her mother "was in agony," and the next day, the "agony of breathing more than she could stand." On December 20 she "had a fearful attack (screaming) . . . Lyle came over and helped me. We gave opiate, she was quiet in 20 minutes, then breath came slowly." Lyle was so alarmed at Mrs. Garrett's condition, and so concerned about Cammie, that he canceled his planned Baton Rouge trip. When conscious, Leudivine sometimes recognized Cammie and sometimes not. In her delirium, sometimes she called Cammie "Mother."[126] On February 3, 1935, Cammie wrote to Carrie: "Mother is just living—had to be given an opiate—suffering so intense. I could not witness the agony. She has stopped taking any nourishment, not even water. So I expect the end is near. Don't try to come. Be too hard on you to witness and you need your strength—life has to go on. I am blessed and have Lyle. Friends have been kindness itself. Try to be calm. This is fate. Coming to us all."[127]

Leudivine died on February 5. In his notification to Carrie, Lyle wrote: "She died peacefully, slipping away in sleep. Aunt Cammie was wonderful, quiet and composed. There was the usual excitement which attends death—no matter how long awaited—people coming, telephone ringing, and the various boys coming from scattered points . . . Aunt Cammie slept, or rested for several hours in preparation for the long ride to Thibodaux today. It was all as painless as possible."[128] Lyle contacted Cammie's friend Robina in Shreveport and had her come to Melrose to stay with Cammie after the funeral. As he explained to Carrie: "Maybe this is a needless thing . . . but I'm afraid that the fatigue and the long strain may cause Aunt Cammie to go to pieces tonight . . . Robina is sensible and quiet and Aunt Cammie likes her."

The funeral was in Thibodaux, where Leudivine was buried next to her husband in St. John Episcopal Church's cemetery. Family friend and newspaper writer Edith Wyatt Moore wrote a lengthy tribute that ran in the *Tensas Gazette* honoring Mrs. Garrett and detailing her love for quilting, plants, and visiting with the Melrose guests. Moore wrote that "the Grand Old Lady of Melrose" was not to be outdone by the younger generation, creating poems, pictures, and lovely quilts.[129] Years later, Ada Jack's cousin and Cammie's friend Lillian Trichel would say: "So much of Mrs. Henry could be seen in her mother, and one of the finest things I saw at Melrose was the care Mrs. Henry gave her Mother, not meddling with her or wrapping her in cotton wool, but letting her do the things she wanted to do and giving these things importance. So many loving daughters make their old mothers a burden by refusing to let them do any work, or by belittling what they manage to do."[130] Cammie was typically stoic about her mother's death, though she confessed to Carrie, "I miss Mother fearfully."[131]

L ife went on at Melrose, and Cammie had too much to do to slow down for long. Ada's visits to Melrose had become so sporadic—and her published work nonexistent—that in 1935, in an attempt to find out what was going on, Cammie wrote a letter to the editors of *Harper's* inquiring about upcoming stories by Ada Jack Carver. In August she received a response from them saying there were not any upcoming stories from the author: "We have published in *Harper's Magazine* a number of stories by Ada Jack Carver, but unhappily we have none now in hand awaiting publication. It must be all of seven or eight years ago since she last appeared in our pages."[132] It was puzzling because Ada had often mentioned specific stories she had in progress and on at least one occasion said one had been accepted by *Harper's,* though it apparently never appeared. Ada's sinus operation evidently had not helped, and in the summer of 1936, Lillian Trichel wrote to Cammie: "Ada and John and David drove out to Boulder, Colorado, where there is some kind of a writer's colony—. . . John was to get them settled and leave them there, and we all hope Ada will be greatly benefitted by the change. She has certainly not been well since that last operation, though I doubt if she would want me to say it."[133]

The next few years would take both Lyle and Carrie away from Melrose in their work; in their absence, Cammie both picked up new threads and continued her focus on her library by collecting diaries and other works for preservation. She was increasingly on the search for relevant books, as seen in many letters between her and Debora Abramson at the Louisiana Library Commission. In July 1936, for example, she expressed interest in Walter L. Fleming's *Louisiana State University, 1860–1896* and, of a more personal interest, the Alice Premble White study of her ancestor Joseph Erwin.[134] She also continued her heavy correspondence with friends across the country, such as artist Edith Fairfax Davenport in Florida, Natchitoches Art Colony founder Irma Sompayrac Willard in New York, and Cora Bristol Nelson in Tennessee, where Bobbie still lived.

Even when Cammie was not actively looking for specific manuscripts for her collection, people would send their work to her for evaluation. Her time was spread so thin that sometimes she was able to read and respond to these requests and sometimes not. The reputation of her Melrose library grew to the point that she received letters more and more frequently from students in search of material for their thesis papers, as if she had time to do research for them. On the back of a letter from a student in 1939 seeking material on Louisiana governor William C. C. Claiborne, Cammie wrote, "I need a private secretary—I shun crowds—but the Library I must help."[135] After Cammie's death, there were found among her papers several unreturned manuscripts and works in progress from various writers. With these many demands on her time, she began to see the need for an assistant to help her, perhaps taking more to heart the advice of Henry Chambers regarding her legacy library. Like the river she lived beside, Cammie knew when to shift course.

5

A Shift in Course, 1937-1940

Why do we have to get done living before we've really learned how to live.
Life has too many unsolvable problems.

— CAMMIE HENRY, letter to François Mignon, January 10, 1939

In July 1937, Lyle Saxon's novel was finally published. *Children of Strangers* had been at least fourteen years in the making, if one counts from the day he stood outside Uncle Israel's cabin and suggested he could write a novel there. Saxon went through three publishers in the process, finally signing with Houghton Mifflin in 1936, after H. C. Kinsey complained that the book was years overdue and requested a return of the advance paid on the novel.[1] Though he had struggled and procrastinated over the years, Lyle did finally sit before his typewriter in that cabin at the edge of the cotton fields and write a work of love, passion, romance, and rejection that captured the caste system of Isle Brevelle.

The novel tells the story of beautiful, young Famie Vidal, a mulatto girl who has an illicit love affair with a white man and bears a son, Joel, on whom she dotes. From the beginning, she is an outcast among the whites because of her mulatto status, and she refuses to associate with the blacks because she feels she is above them; by the end of the novel, she is also an outcast among her own people. The characters in *Children of Strangers* were based on people Lyle knew from Isle Brevelle. For example, Guy and Adelaide Randolph, owners of "Yucca Plantation," are drawn from Cammie and John Henry, confirmed by Joe Henry in James Thomas's biography of Saxon; and Mr. Paul, Guy Randolph's brother, is based on Lyle himself. Gravely ill, Mr. Paul has returned to Yucca to die, and he befriends Henry Tyler, a kind, gentle black man whom Famie marries at the end of the novel, after the mulatto society will have nothing to do with her. Famie's cousin Numa is the man her family expected her to marry. All of these characters and places in the novel were drawn from the Cane River area.

By the time *Children of Strangers* was released, Lyle was involved in his Works Progress Administration (WPA) employment. In an interview with Jo Thompson of the *Baton Rouge State Times,* he modestly said: "I've been working so darned hard on this WPA writers project that the book fairly slipped out. There it was, published one day and I could hardly realize it until the next day."[2] It is an astounding quotation, given his long struggle with the novel. It is as if the book was nothing more than a litter of kittens that just appeared under the cabin one day. The statement belies the intense revising Saxon did on the novel; it was at one time thousands of pages, and he edited it heavily, telling Thompson that he "sometimes boiled a page of the original manuscript down to one sentence." There were even rumors that the novel was so long in coming because some of the Melrose help had "inadvertently destroyed" part of the manuscript, but that hearsay was unfounded.[3]

The long anticipated novel received good reviews and sold very well, especially in Louisiana. By this time in his career, Saxon had been dubbed "Mr. New Orleans," and he was at the peak of his fame as a writer. In the *New Orleans Times-Picayune*'s list of top-selling books in New Orleans for the weeks of September 12 and 19, 1937, *Children of Strangers* sold more copies locally than John Steinbeck's *Of Mice and Men.* Saxon's book sold well nationally but was not in the overall top ten for 1937, first place going to *Gone with the Wind* by Margaret Mitchell. *Of Mice and Men* outsold Saxon's novel nationally, as did *Northwest Passage* by Kenneth Roberts. Cammie clipped the best-seller lists documenting both regional and national sales from the *Times-Picayune* and pasted them into her scrapbooks. As she had in the past, she retained a press clipping service to collect articles from around the country so she could monitor and save reviews of the novel. George Stevens, writing for the *Saturday Review,* said: "The mulattoes offer such abundant material for 'local color' writing that the novel is a little slow in getting under way; and elsewhere the descriptions sometimes tend to interrupt rather than illuminate the story . . . Possibly it started out to be a different kind of novel, in which the community itself was the dominant theme, until Famie ran away with it. In any case, the point is a minor one, because Famie has run away with it successfully. If the novel is not sociological, neither is it primarily regional. The mulatto community is unique, and interesting in itself, but it is the character of Famie, developed in all dimensions and human terms, that gives the book its poignancy and its considerable distinction."[4]

Edward Larocque Tinker, writing for the New York Times Book Review in July 1937, noted the multiple postponements of the novel and concluded this was for the best. It "bears the evidence of years of careful work," he said. "The book has benefitted greatly from this long period of gestation and presents a seemingly effortless spontaneity and finish that is the hallmark of literary craftsmanship." Tinker said Lyle presented his material as "an artist, not as a propagandist."[5]

Cammie's clipping service sent reviews and articles from every major, and many minor, newspapers and magazines from all over the country; many of these articles had the same photo of Cane River mulatto Zeline Roque taken by Tulane University professor Frans Blom alongside the review. The date of the photo is not known, but Blom was an associate of Lyle Saxon and was depicted in William Spratling and William Faulkner's book *Sherwood Anderson and Other Famous Creoles*. He was a Danish-born archaeologist who focused on Mayan studies, and during his tenure at Tulane (1924–41), he remained involved with the Arts and Crafts Club in New Orleans, occasionally showing photographs and movies from his Yucatán travels.[6] Blom visited Natchitoches at least once in 1931 to lecture at the Normal School, and it likely that he met Cammie Henry. It is certainly through Saxon that Blom met and photographed Zeline Roque on Cane River.[7]

There was Zeline in the *Toledo (Ohio) Blade* and the *Los Angeles Times*; Zeline with her headscarf, head cocked to one side, staring into the camera with a hint of a smile on her face; Zeline leaning on her cane, one elbow perched on the rail of her steps, now a familiar face splashed across newspapers nationwide. The little Cane River community of Isle Brevelle, now made famous in Lyle's novel, was a curiosity.

Lyle was nervous about how the novel would be received on Isle Brevelle, and perhaps this concern led in part to his great care in finishing the novel. Carrie Dormon, in her 1937 diary, wrote that Lyle "sent a copy to Father Bumgartner, the priest of Isle Brevelle" and "felt a little uneasy as to how it might sit" with him. It was, after all, a rather lusty tale of forbidden love and romance. "Miss Cammie was inspired with the idea that we go see [Father Bumgartner]. We did. He and Lyle had quite a chat, and we left the Rev. Father all smiles."[8]

Carrie noted in her diary that Lyle "is in fine spirits over his book . . . and well he might be. It is a splendid piece of work. He has done a difficult piece of writing with delicacy, yet strength . . . he is truly gifted."[9] Even in this very private forum, there is no sign of jealousy or resentment from Carrie that Lyle had published his novel while she had not published a single word of her own fiction. Cammie sent an autographed copy of the novel to Cora Bristol Nelson, who was delighted with the book.[10] Lyle's friend author Julia Peterkin sent a note to Cammie declaring *Children of Strangers* to be "a good piece of work" that will "give pleasure to many readers."[11] There is no record of Ada Jack Carver's reaction to the novel.

Lyle did not have much time to rest on his laurels; he would soon be in Baton Rouge and in Washington, D.C., working on the WPA's *Louisiana: A Guide to the State*. And in February 1938 Carrie began a job in Pineville, landscaping the new hospital. She wrote to Cammie: "Don't faint—but I've accepted a steady job—that is, I know it will last at least 18 months. When you know what it is, I think you will approve . . . when the question of planting the grounds of the new charity hospital (Pineville) came up, [A.R. Johnson] said he wanted it in native things, and wanted *me* to take charge! It is 40 acres, beautiful, rolling grounds. And he promises I can have a *free hand*."[12]

Carrie's plate was now full. She had been appointed in 1935 as the only woman to Roosevelt's DeSoto Commission—a prestigious committee formed to commemorate the four hundredth anniversary of Hernando DeSoto's exploration through the southeastern United States (1539–43). She had accepted begrudgingly, though, as government bureaucrats irritated her. She traveled to Washington, D.C., and to Tampa, Florida, for meetings of the commission to study the route of DeSoto's expedition.[13] Carrie likely resented the time that this appointment required as she preferred to be in the woods and working with her plants, but she also had a strong interest in seeing that the committee's work was done right, recognizing the value of the anniversary.

I t was also in 1938 when Cammie first met François Mignon. On a trip from New York to Louisiana with his companion, Christian Belle, Mignon met Lyle at a party at Roark and Mary Bradford's house in New Orleans. Lyle must have known that Cammie and François would have plenty in common, and he invited Mignon and Belle to visit him at Melrose, completely unaware that they had already been invited by Cammie through her friend Edith Wyatt Moore when they passed through Natchez, Mississippi. Mignon was at a transition point in his life. His eyesight was weak and failing, and with the outbreak of war in Europe, he contended that his work in foreign trade was diminished. Belle was being transferred from New York to Puerto Rico through his work in the French diplomatic corps, and Mignon found it necessary to consider a move as well.

François Mignon was a man who had reinvented himself, contriving an identity based on falsehood. Born Frank VerNooy Mineah in Cortland, New York, in 1899, he attended Columbia University briefly and later worked for B. Westermann Company, an international bookstore in New York City, from 1932 to 1939. A Francophile, he adopted the pseudonym François Mignon in the mid-1930s, although no one knows exactly why. By the time he arrived at Melrose, he was no longer a native of New York who had attended school in Cortland but had convinced even his closest friends that he grew up in the Île de France and was educated at the Sorbonne. No evidence of either exists. In uncovering the truth of Mignon's birth, Oliver Ford points out that regardless of his fictitious beginnings, Mignon ultimately embodied the life he created and was a well-liked and popular man in his adopted Louisiana. "There are ample testaments to his kindness and gentleness," Ford writes.[14]

Cammie and François shared interests in preservation and in Natchez; she also saw in him someone who could help her with her library and her increasingly voluminous correspondence. With Lyle mostly absent from Melrose, Cammie became more and more determined that François return to stay, as if on some subconscious level she knew that Lyle's time of residence at the plantation was ending and François could perhaps fill that void. In September 1938 she wrote to him upon his return to New York: "I have you in my mind constantly, not 'ships that pass in the night' but congenial contact for a life time—I pray. I did not see half enough of either of you—so anxious you get the valuable association of Lyle Saxon—he who is steeped in Louisiana lore—now the thing to do is to come back to Melrose your first opportunity—that I may know you . . . my mind will not be at rest till you do this."[15]

Cammie was quite concerned for François's health and knew that once Belle left the country, he would be quite helpless and alone as his vision deteriorated.

In the same letter she urged him to go to the Mayo Clinic to see what could be done about his failing eyesight. She signed this letter formally, "With affection, Carmelite Garrett Henry," while two months later she would sign her letters to him "Aunt Cammie." In July 1939 she outlined her conditions to stay at Melrose, assuring him that her customary rules would not be a problem for him: "There is just one requirement & you need never be told that . . . Dr. Rivers of Mayos asked what it cost to stay at Melrose—I ans[wered] that you produce—we shelter—no idleness—that's easy for you—I do not rent the cabins—friends who need quiet to work occupy them—that's all the charge."[16]

In October 1938, still a year before François would return to Louisiana, Cammie and Lyle planned a trip to Natchez together. She was certain that if she could get Lyle to Natchez, "it won't take him long to feel the pulse of that city."[17] With his credentials as "Mr. New Orleans" and his nonfiction works *Fabulous New Orleans* and *Old Louisiana* to his credit, Cammie hoped he would turn his attentions to Natchez with her; he could "ferret out" what could be done in the way of preservation and engage a plan of action just as he had with the Vieux Carré. The trip fell through when Cammie stepped on a nail and was unable to go. She encouraged Lyle to proceed without her, but he did not.

François, on the other hand, was quite interested in Natchez and the early preservation efforts under way there. Cammie found in him "a congenial soul," much like she had with Lyle, Carrie, and even Ada. She was working on a scrapbook of Louisiana plantations in November 1938, when she urged François to return to Melrose: "I am working on a volume of La. Antebellum plantations . . . things fast passing away—the government is doing a splendid piece of work on this period—measuring—photographing old homes—so clearly that future generations can rebuild. Will anyone want to?"[18]

Cammie likely saw a sort of parallel between Natchez and Isle Brevelle in that both had been isolated from the outside world. In 1938 work began on the bridge over the Mississippi River connecting Natchez and Vidalia, Louisiana. Cammie was afraid that this bridge, and this new accessibility to Natchez, would bring too much change and too much progress to the town, writing to François: "I sit and see the disintegration of Isle Brevelle . . . I can't hold back time— and what's called 'progress' is to me—often destruction. This generation won't even care or know how to save legend and tradition."[19] Indeed, the fear was that the bridge near St. Augustine church connecting Isle Brevelle with the highway near Melrose could be contributing to the erosion of its culture by making it too open to outsiders. In January 1939 Cammie wrote to François: "When that bridge is completed across the Mississippi—there will be a fearful traffic through Natchez—the place was saved because it had neither highway nor railroad. This Natchez trace may all be fine but it will reduce values—of worthwhile things—wipe out legend and traditions   . . . I hate to see it happen before a real plan of preservation is under way."[20] Development of the Natchez Trace Parkway began in the 1930s as one of the Civilian Conservation Corps projects. It followed the original Natchez Trace trail that dated back to the Natchez Indians, who, along with the bison and other wild game, left a four hundred–mile trail through the woods that ran from Mississippi to Tennessee. As the federal government made plans to lay down asphalt, Cammie's anxiety grew over the idea of change without a plan for preservation. At some point she likely felt confident that she had preserved and documented as much of Isle Brevelle and antebellum Louisiana as she could through her scrapbooks and her collection of buildings, anecdotes, and memorabilia. Branching out to antebellum Natchez, which she saw as threatened, may have made sense as the next step.

Her new friendship with François Mignon reinvigorated Cammie. In December 1938 she wrote to him: "Your letter did my soul good—we think so much alike. Just friendship makes life—and it means loyalty, companionship, congeniality . . . this passing year has enriched me greatly—did it not give me you and Christian?—and I will hold to you both? Grow old along with me—. . . friendship like wine, ripens with age—. Oh, so glad I found you—I feel like you are made for Louisiana—preserve Louisiana lore—the part most people miss."[21] After years and years of tourists, writers, and students of history coming to avail themselves of her library, Cammie had reached a point in her life when, as she once told Lyle, she did not want to meet any more "new people." Lyle's response had been "'Oh yes, but you are, dear—for one has to meet one hundred to get one really worthwhile'—so I have taken back my rash assertion—and made my bow to Mr. Saxon—for being right as usual."[22] This conversation made such an impression on her that she recounted it in more than one letter to François.

In February 1939 another Natchez trip was planned with Lyle, but it, too, fell through because he was tied up on WPA business and the weather was bad.[23] As Lyle was working on the Louisiana guidebook, he consulted with Carrie about including Briarwood in one of the motor tours: "Our tour passes from Saline to Chestnut just at your front door, and I have been wondering whether you would prefer to have Briarwood included as a point of interest in the tour or not."[24] He knew how intensely private Carrie could be, yet including Briarwood in the tour, he pointed out, "gives us an opportunity to mention your *Louisiana Wild Flowers*" book. Carrie decided to decline, primarily because the place needed some work but also because she and Virginia were seldom at home. She suggested Lyle include the nearby Drake saltworks instead. He also attempted to get Ada involved with the WPA project; she was appointed to the advisory committee, but there is little evidence that she participated to any great degree.[25]

Cammie had begun to look toward Natchez, but in January 1939, when she learned that her ancestral home, Shady Grove, was to be demolished, once again her preservation instincts resurfaced. Shady Grove had been used as a high school in Iberville Parish for twenty years. Cammie wrote to the principal of the school and attempted to purchase the colored glass door and sidelights "as a memento of my grandfather's early home," adding, "I remember very pleasantly a visit to Shady Grove a few years ago when Mr. Lyle Saxon went through the old buildings and admired this work that my grandfather had wrought."[26] Whether her request was granted is not known.

Early 1939 brought Lyle back to Melrose, however briefly. He had completed work on the tour section of the Louisiana guidebook, and he returned to his cabin, where he began a slow, downward spiral and a heavy drinking binge, sitting up late into the night imbibing whiskey. On March 1 he had surgery for a ruptured appendix after collapsing in his New Orleans hotel room. His close friend Edward Dreyer wrote to Robina Denholm on March 13 to tell her that Lyle "has now come out of the woods and all the doctors say that he is getting along alright" following the operation. "He was in the operating room for some two hours in as much as his appendix was misplaced and they had a hard time finding it and when they did find it had a hard time getting it out." Dreyer and another friend alternated twelve-hour shifts sitting with Lyle during his recovery for the first eleven days, "when he had more tubes going in and out of him than any one person could regulate . . . he had delirium and insisted that he had to catch a train for Chicago or go to a party" and had to be subdued. Finally, he came through it, his appetite returned, and he was allowed to eat, "including chicken, steak, etc., and he has a highball before every meal."[27] Lyle stayed in

New Orleans to recover and felt well enough by early 1940 to dress as a giant rabbit for Mardi Gras.

Cammie continued her heavy correspondence with François through most of 1939, until finally he arrived at Melrose in the fall. He was forty years old at the time, twenty-eight years younger than Cammie. Robina Denholm and François had become friends through their mutual association with Melrose and often wrote letters to each other. In August 1939, for example, Robina wrote a letter that certainly must have made François long for the beauty and camaraderie of Melrose: "It is Sunday at Melrose—dinner is just over—a grand dinner which started with gumbo and then fried chicken—piled high on a huge platter—then all the rest and finally black, black coffee. It was a family gathering and we had a gay time, the table was set in the summer dining room and there was a ceiling fan over us . . . This is Lyle's typewriter and I find it difficult to make a go of it— perhaps because it seems stiff from long periods of disuse. Anyway, his cabin is lovely, the breeze is grand, and there is a delightful quietness that I love."[28]

François's visit to Melrose was initially intended to be for a few weeks, but he quickly became indispensable to Cammie and stayed for thirty years. He and Cammie set to work almost immediately on her library, for which she began to investigate once again the possibility of getting a catalog printed. In August 1939 Cammie wrote to John Andreassen, regional director of the Historical Records Survey (HRS) in Louisiana, to ensure that she be put on their mailing list to receive their future publications. The HRS, part of the WPA, was purposed with indexing the records of historic societies, courthouse records, manuscript collections, and other collections that would preserve and centralize this type of information. Cammie continued writing to publishers, booksellers, historical societies, and libraries for books pertinent to her interests in Louisiana history. Her correspondence with Debora Abramson of the Louisiana Library Commission became so frequent that Abramson began to address her letters to "Dear Aunt Cammie" rather than a more formal address previously used. François began cataloging the scrapbooks, typing up indexes for the ones of historic import, which he then placed inside the front covers.

Along with Carrie's and Lyle's diaries and Cammie's 1934 diary, the journal of François Mignon is another treasured record of daily life at Melrose. Beginning upon his arrival on October 28, 1939, the journal describes the general and the mundane activities from François's perspective, however much he can be trusted. It is rich in detail from plantation routines like morning coffee to jaunts to Natchez, sometimes with Cammie and sometimes without. Lengthy, typed daily entries are often filled with eloquent, descriptive prose. It must be suggested that François's recollections be taken "with a grain of salt," as it has been documented that he is sometimes incorrect in what he remembers, sometimes embellishes, and sometimes is too inventive. Some of his tales were simply fantasy or elaborated. François's fabrications extended further than his own origins: it was he who named Lyle's cabin "Yucca" and who dubbed the building Cammie called the mushroom house "the African House."[29]

Robina Denholm drove François to Melrose once he reached Shreveport from New York. When they arrived, Cammie greeted him warmly, and he observed that she was "looking so good and so wholesome in her neat white waist, black skirt and her luxuriant white hair."[30] Cammie had fixed up the old detached kitchen, which had been in use as a studio, for François to live in, which delighted him. It was situated immediately behind the main house but in front of, and to the right of, Lyle's cabin. It was at one time used as a blacksmith shop, François claimed, but Cammie had "metamorphosed it into a beautiful cabin, with the main room two storeys high, and lighted by a wonderous Louis XIV window with beautiful fan light which soared toward the ceiling. There were vases of beautiful flowers—red and yellow dahlias of generous proportions, another with pink dahlias with some wonderful blue flowers that looked a little like lilacs." This was the room Alberta Kinsey had used as a studio in previous years.

François settled comfortably into what he often referred to as his "maisonette" and quickly established a daily routine: taking morning coffee delivered by Frank Moran, who would stoke the fire and return shortly after with a breakfast tray; going over the mail and having coffee with Cammie around ten; and often taking long walks around Isle Brevelle. François developed friendships with the community's residents, including Zeline Roque, whom he visited regularly, often bringing her supplies from Melrose that Cammie had gathered. His long walks "in the big road," as François called the River Road, sometimes included catching a lift with a passing automobile or school bus, and he thus quickly met all of the neighbors. François appeared fascinated with the whole process of raising cotton, and his journal reflects numerous references to the "lushness of greenery and opulence" of the growing plants and the operations of the gin, where he would go "watch the miracle of mounds of cotton being swept from the trucks, through the gin, and out into 500 pound bales."[31] At night he would walk through the fields and admire the bolls of cotton, "soft and shimmering in the moonlight."

Cammie and François soon began a series of jaunts to Natchez over the next few years. Sometimes these were day trips. Their route to Natchez was about one hundred miles, and it took only a few hours by car—over mostly dirt roads and on a ferry—to get there. Other times they stayed a few days. On occasion François went either alone or with Robina. He recorded lengthy, detailed descriptions of one plantation home after another in his journal. Before one of these trips, in November 1939, Cammie left instructions with the servants and "admonitions not to burn the house down," and then she and François were off to Natchez. In his journal François wrote, "At Ferriday we noticed the vast construction on the new emplacements for the new bridge which leads to Natchez, —and when finished, will eliminate the isolation and much of the almost forgotten ante-bellum charm."[32] They took the ferry across around noon and headed into town to Edith Wyatt Moore's home.

Moore was born in Georgia and educated in Tennessee and wrote for a period of time for the *Chattanooga News*. Upon her marriage in 1919, she moved to Natchez and wrote articles for the *Natchez Democrat*. She founded the Old Natchez District Historical Society and was the go-to authority on the city's past. Eventually, she was made district supervisor for the Federal Writers' Project guide series in Mississippi, a role Lyle Saxon assumed in Louisiana. In 1958 Moore published *Natchez under the Hill*, a history of the district around the original Natchez port and the beginning of the Natchez Trace. She and Cammie had become good friends, and she was more than qualified to introduce her to the historic sites in Natchez.

The trip was filled with tours of plantation homes that Cammie wanted to visit and document for her scrapbooks. At Foster's Mound, François tried to imagine what plantation life had been like as they stood on the gallery of the old house, there since the 1790s, with its blue painted ceiling (a favorite color on southern plantations because it reportedly repels wasps and dirt daubers), and admired the surrounding countryside.[33] They also visited the site of Windsor Plantation, which had burned in 1890, and viewed the ancient live oaks. The original iron steps of Windsor had been salvaged and installed at Oakland Chapel at Alcorn College, which they saw as well. The return trip to Melrose provided an opportunity to stop at the Cottage Plantation, just north of St. Francisville, to visit Mrs. Louise Butler, an old friend of Cammie. François's journal details the meals they ate and records conversation threads. While rather uneventful, the trip was typical of many that Cammie made with François to explore Natchez, tour the old plantations, and fret over the march of progress that she feared would ruin the city.

Back at Melrose, Cammie received a phone call from Lyle with news that his friend the newspaperman and drama critic Alexander Woollcott, who at that time wrote for the *New Yorker* magazine, was to arrive at the plantation for the night. Woollcott brought his large black French poodle and over dinner told them all about the Dionne quintuplets, with whom he had just been working on a short documentary, *Five Times Five*.[34] After dinner Cammie brought out her McAlpin stencil, which then led to a lengthy discussion of Harriet Beecher Stowe and her career-ending decision to write about Lord Byron and his relationship with his half-sister. After the dinner conversation, Woollcott retired to Lyle's cabin to spend the night before his departure the next day. Upon his return to Dallas, he autographed some books and sent them to Cammie for her library.[35]

Woollcott was one of several prominent writers who visited Melrose via an association with Lyle Saxon. It has long been rumored that Lyle's friendship with William Faulkner and William Spratling in their early New Orleans days brought Faulkner to the plantation for a visit. James Thomas interviewed Joe Henry for his biography of Saxon and quotes Joe as being certain that Faulkner had been there, though no other supporting evidence has been found. Even if Faulkner did make a passing visit to Melrose, he never spent any significant time there.

I n his journal François often wrote of walking up the road to the saloon with Lyle whenever Saxon returned to Melrose. The two would also frequently pass time drinking, discussing books, or visiting neighbors around Isle Brevelle. In November 1939 François noted that Lyle arrived on the evening train at Derry, and after greeting Cammie, who was suffering from pleurisy and thus retiring early, he went with François to his cabin, where houseman Frank Moran had a fire blazing for them. They stayed up until the early hours of the morning, catching up on friends and other gossip; Lyle amused François with stories about Grace King, politics, and his then legendary appendectomy. By 3:00 a.m. the drinks had run out, and the two called it a night.

On the same visit by Lyle, he and François paid a call to Zeline Roque, now eighty-five years old. Lyle gave her some money and promised to send her some medicine for her rheumatism. On their way back, walking along the river road, the two men stopped at the saloon, which was segregated under Jim Crow laws of the time: "A low, white-washed building, set in one of the triangles formed by the three roads converging at the Melrose bridge. From its low gallery, one has a marvelous sweep of the river . . . Inside, the saloon is divided by a partition which cuts the place in two, the bar at the rear marking the end of the partition. With this arrangement, the white customer enters by the door at the right end of the gallery and the colored at the left door, so that while neither white nor black can see the other while they stand at the bar, one barman can see all his customers and take care of their wants with dispatch."[36]

The two men took their drinks out onto the gallery to watch the river while they talked, a fairly regular routine of theirs. A couple of weeks later, Lyle was again at Melrose, and they repeated the visit to Zeline, followed by drinks on the saloon gallery. They would drink most of the afternoon, often with Lyle buying drinks for everyone in the place, returning home to Melrose in time for one more highball before supper. The nightly pattern included conversation or read-ing before the Franklin stove in Cammie's bedroom—it did not necessarily have to be lit—and then François and Lyle would grab some ice from downstairs be-fore heading over to the cabin for more drinks and conversation.

François enjoyed Lyle's visits to Melrose and loved hearing his stories and anecdotes about the luminaries with whom he associated, but it is unlikely that Mignon drank to the extent that Saxon did. There's no proof that François drank regularly in Saxon's absence. While Lyle was probably a functional alcoholic, anecdotal evidence indicates that Cammie did not allow alcohol in the house, nor did she drink, outside of perhaps a glass of port or other wine at Christmas. She certainly must have known that Lyle had liquor in the cabin and that he and François visited the saloon, but she showed no concern and did not appear to judge anyone who did drink socially.[37]

M elrose was a working plantation, and running the household was a full-time occupation, part of which included managing the employees. Cammie's su-pervisory techniques were sometimes successful, sometimes comical, and some-times frustrating. The occasion arose in November 1939, for example, when she needed to hire a new cook. Cammie interviewed Elmer, a woman who had come into the plantation store and told J.H. she was interested in the job. When Elmer came to the house for her interview, Cammie asked her: "Now do you really want the job as cook for me? Because I don't want anybody to work here who doesn't want to. I know the people who work here and I like them, and I want them to want to work here or not work here at all." Elmer affirmed that she did indeed want the job, though she did not know how to cook. "Well," Cammie said, "that's alright. I know you can learn and I'll teach you." So Elmer was hired and moved to Melrose with her three children, her hens, and her pigs.[38]

Elmer replaced Clémence, known today as the artist Clementine Hunter. Clémence was to teach Elmer how to cook, but it seems that she found it easier to do the work herself for the most part rather than instruct the hapless Elmer.[39] One of Elmer's duties was to set the table, a task that was not always successfully completed. François recalled supper one evening when the table was set with soupspoons they did not need and "a scarcity of bread plates" they did need: "Aunt Cammie had especially requested toast be served for Dan, but no toast appeared. Asked if she had forgotten it, Elmer said she hadn't. She didn't serve any toast because she didn't know what toast was or how one finds it. Aunt Cammie was marvelous. She merely said, 'That's alright Elmer,—you'll learn and then it will be easy.'"[40]

Cammie seemed to take the dinner table mishaps in stride, always finding something humorous about the situation and often gently suggesting a way to avoid repetition of whatever mistake had been made. François wrote in his journal that he "never once . . . heard her raise her voice" with the servants. The only thing bad about mistakes, she said, was if you failed to learn from them.[41] With Thanksgiving dinner looming, Frank Moran was as nervous as he could be about serving family and friends with Elmer in the kitchen. Making light of the situation, he teased Cammie that he would get himself a dress to put on and serve dinner "and then I'll be in the place you need me!" Cammie laughed, told him he would do no such thing, and assured him that everything would be fine.[42]

Try as she might, Elmer struggled with even the simplest tasks; she had been on the job over two months when François lamented that "poor Elmer can't even make coffee twice the same—which perhaps is fortunate since the last time always seems worse than the one before last." Both Cammie and Frank showed Elmer how to measure the grounds several times and how much water to add, but Elmer's coffee remained a work in progress.

One important contribution of François was promoting the Creole self-taught folk artist Clementine Hunter, who had worked for Cammie Henry as a field hand, then cook, and finally house servant. Born Clémence Reuben in 1886 or 1887, she grew up working on the grounds of Melrose. Her father, John Reuben, was hired by Cammie's husband to work on the plantation when Clémence was a teenager. In recorded interviews Clementine says she liked working in the fields picking cotton and was proud of how much she could pick each day, and she liked picking pecans. All of this would later inform her art. The story goes that she found some discarded paint tubes left behind by Alberta Kinsey and told François she would "mark" a picture for him. After that Alberta often gave paint to Clementine. François championed Clementine's painting through the years and helped her gain recognition for her work.

The first instance François discussed Clémence at any length in his 1939 journal was mid-December. At that point she was no longer the cook but still worked in the house for Cammie. She was also there to step in for Elmer when needed or if extra company came and Elmer needed a more expert hand. François described Clémence as "thin as a rail and looks about 35 but they say she is in her sixties."[43] Clémence could turn out "a beautiful dinner, and with a wave of the dishcloth the dishwashing is out of the way, and Clémence is on her way to some other undertaking that she likes better."

Clémence was a fine seamstress and quilter and also made dolls. Cammie taught her and some of the other employees to spin, but Hunter was not particularly fond of doing it. "No matter what cajoling or persuasion Aunt Cammie may use on her, Clemence simply won't spin cotton for Aunt Cammie because she simply does not like to spin," François said. Hunter was painting some by that time and promised to bring François one of her pictures to examine. He wrote that she "works in oils, using the bottom of a cardboard box or the side of a Kirkman's soap box as a canvas. And they say her creations are something, whether they be still life, landscapes, or portraits. The one she is going to bring me is a scene in a sick room, a man seated in a chair with a woman nursing him. When I ask her about her models, she explains that she never needed any. Sometimes at night, she explained, she has a vision, and getting up she searches around until she finds something to paint on and then just goes ahead with the paints that Albert McKensie [sic] has given her."[44]

Cammie would likely have been aware that Clémence was painting, if for no other reason than that as closely as François worked with Cammie, he would have told her about Clémence's artistic efforts. However, there is no real evidence that Cammie ever appreciated Clémence's art in the same way that she did Kinsey's. She would have seen Clémence simply as the cook or the hired help.

Cammie Henry must be viewed through the lens of her historical period and upbringing. The caste system of Cane River was observed even to the point that it was a major theme in Saxon's *Children of Strangers*. The mulattoes saw themselves as superior to blacks, and the whites felt superior to the mulattoes. While everyone at Melrose got along, in general blacks and mulattoes did not take meals together. For Cammie to have invited a black artist to sit down at her dinner table and to be served by her hired staff would have been unthinkable to her. It would have never crossed Cammie's mind to treat Clementine Hunter at the same level as Alberta Kinsey, for example, not only because of race but because Cammie kept the demarcation between friends and hired help clear.

Holidays at Melrose always brought family to the table, at least those who were able to get there. Christmas dinner in 1939 consisted of gumbo, turkey with oyster dressing, cauliflower, roast pork, potatoes, greens, biscuits, cornbread, cranberry sauce, and an assortment of jellies and pickles. J.H. had spent the morning at the plantation store, which was always open on Christmas morning, with the counters loaded with spirits for any of the employees who wanted to stop by: "As each was welcomed, he was asked what he would like for Christmas, some choosing overalls, some fireworks, some tobacco—and some of these who had been on the place longer were given suits."[45] The day continued for Cammie and François with fruitcake and coffee and later a cold supper of red wine and cold turkey followed by an evening opening Christmas mail.

As another decade drew to a close, Cammie's inner circle of congenial souls was less cohesive than in previous years. Carrie had wrapped up her work on the DeSoto Commission, Lyle was in New Orleans resting and recuperating from his health troubles, and Ada had all but disappeared into her own world. But with François in residence, Cammie took things in stride, focusing on Natchez, her library, and other new pursuits.

6

Cultivating the Legacy, 1940–1948

I sat on my gallery for a while, enjoying the close of this
warm summer's day which seems more like May than November.

—FRANÇOIS MIGNON, journal entry, November 16, 1939

I n 1939 another world war erupted, and its reach extended to the Cane River region. The United States War Department, in assessing its readiness to take on a more mechanized and advanced German military, determined that the American army needed to modernize; the area of Sabine Parish in Louisiana proved to be the perfect training ground for testing new tactics and operations. The Louisiana Maneuvers began on a small scale in May 1940 and continued until 1941, involving 400,000 soldiers.

Cammie's eldest son, Stephen, returned to Louisiana to participate, enabling her to see him a little more often.[1] Soldiers involved in the war games were present throughout the Cane River region and interacted frequently with the locals. Newspapers across the state carried updates on the Red and Blue armies, as residents fed the soldiers fried chicken and water from their wells.[2] François Mignon recorded in his journal that there were "large Army trucks parked" at the nearby saloon that he and Lyle frequented and "a couple of dozen privates and officers in and about the saloon and in and about the Cane River—some of them diving off the bridge and some splashing about the shore while the more serious ones were fishing along the banks."[3] When Cammie and François would travel to Natchez, they saw soldiers encamped around Grand Ecore just north of Natchitoches on the bluffs over the Red River. This military activity brought the world a little closer to rural Melrose and provided topics for conversation at the dinner table.

Cammie, Robina, and François set out from Melrose early one morning in spring 1940 for a drive to Natchez to visit friend Mary Lambdin and to tour various historic houses, such as Cherokee Plantation and Edgewood. After returning home, Cammie wrote to Carrie, telling her about "a perfect visit." But Carrie appeared not to share Cammie's passion for plantation homes, being enamored more with the outdoors than with material things like houses and antiques. Cammie and Lambdin practiced weaving as a hobby, and Cammie had set up a loom at Melrose for her use when she came to visit. Cammie's archives include voluminous correspondence between the two about looms, weaving patterns, and discussion of current projects.

When Cammie and François returned from one of their trips, they spent days catching up on correspondence that had come in during their absence, and Cammie had much work to do in her garden as well. She may have brought back something that needed to be planted, or she would need to tend to weeding and pruning chores. Typically, J.H. would relate any plantation news, which on one occasion included a minor theft at the store, ultimately causing François to ponder "the position of the colored man."

As he saw it, despite their freedom, blacks were still basically financially tied to the plantation system and therefore all too often unable to truly leave it.[4] In his 1940 journal, François quoted J.H., who told him that on Melrose, plantation workers made about $3.75 per week. François wrote, "J.H. says they only need about two dollars and seventy-five cents for groceries, meats, etc., leaving seventy-five cents for them to buy their bottle of wine on Saturday night."[5] In truth, on most plantations across the South, workers were given a line of credit, or scrip, in the plantation store and a very small salary or a plot to farm. Once their monthly debt to the plantation store was settled, there would not be much money left over. There did not seem to be an obvious solution to the "complexity of the human problem," François mused. Jim Crow prevailed in Louisiana at this time, and blacks dealt with inferior schools and segregated churches, housing, and parks. The concerted struggle for civil rights lay ahead in the 1950s, and the Civil Rights Act would not come until 1964.

The hired staff at Melrose seemed to be satisfied to work there, judging from the loyalty they gave to Cammie and she to them. In 1936, for example, former Melrose employee Henry Tyler wrote a letter to Cammie after failing to find a "four post wooden bed" for her. Tyler apologized, offered a washstand instead, and closed with "I remain your friend."[6] In his journal François Mignon commented more than once on Cammie's good nature and humor in dealing with the servants. One afternoon Henry Hertzog was untangling a snarl of thread in the loom and asked Cammie for a pin to work the tangle loose. She pulled off the gold brooch she always wore that had belonged to her grandmother and told Henry that she was never without it. She explained she had used it just that morning to work a thorn out of her finger and had even had to pin up the jaws of Uncle Israel when he died, thus sending Henry and Cammie into a hilarious discussion of ghosts and spirits and her vow to haunt him after she died. "It was marvelous to see how the Madam maintains her relationship with the servants," François wrote, "with such a nicety of balance between humor and seriousness, with a slipping of one mood into the other with such adroitness."[7] The affection of the staff for Cammie was also on display in their large turnout for her funeral, both blacks and whites filling the downstairs library at Melrose, no segregation enforced there.

In 1940 the Lemee House on Jefferson Street in Natchitoches was purchased by the city for preservation and as a meeting place for the Women's Club. Built in 1837, it had been a family residence and a branch of the Union Bank of New Orleans through the years. In the earlier fervor of Cammie's transport and restoration of old structures from around the parish to Melrose, Lemee House had been one that she had eyed. J.H., though, had dissuaded her from purchasing it, telling Cammie if she was interested in the house as an antique, that was one thing, but as an investment it simply was not advisable. François suggested that Cammie always regretted her decision not to purchase Lemee House. Today it is the headquarters of the Association for the Preservation of Historic Natchitoches, the same organization that now owns and operates Melrose.[8] This arrangement would certainly please Cammie.

At this point in her life, Cammie seemed less interested in obtaining buildings or cabins and more focused on her library. With François there to help her, she made an effort to obtain local diaries and other specific works reflecting nineteenth-century plantation life. In search of specific volumes, she increased her correspondence with university libraries throughout the state, including the

Louisiana State University Libraries and the Louisiana Library Commission. She was interested in diaries such as B. L. C. Wailes's and Eliza Magruder's, both documenting plantation life of the antebellum South, specifically areas around Natchez. In January 1940 Cammie wrote to the Library Commission in search of Walton R. Patrick's 1937 thesis, which examined literature found in Louisiana plantation homes prior to the Civil War.[9] The focus of her collection during this period seems to have been primarily on unpublished, truly unique works that would separate her library from any other.

Preservation of a vanishing past and recording the present was of concern not only to Cammie but also to photographers and documentarians who worked with the Historic American Buildings Survey (HABS), the Resettlement Administration (RA), and the Farm Security Administration (FSA) as part of Roosevelt's New Deal plan to bring relief to rural America during the Depression. Cammie was suffering through June 1940 with another malaria outbreak and, along with Robina, had decided to go to Hot Springs for three weeks; therefore, she was absent from Melrose in June when Marion Post Wolcott, one of Roy Stryker's FSA photographers, came to photograph the Cane River region.

Wolcott's visit is documented in François Mignon's journal: Wolcott arrived at the plantation on July 1 after photographing other locations in the area. She was greeted by J.H. and his wife, Celeste, who then brought her to François, where they "mapped out subjects" for her to photograph.[10] Rains over the next few days delayed the photography project, but on July 6 Wolcott returned to Melrose. François took her first to the cemetery at St. Augustine Catholic Church and then to Magnolia Plantation and to Derry.[11] Rains continued to interfere with the project, and while François wanted to take Wolcott to photograph Joe and Zeline Roque, the muddy roads made that impossible. It was not until July 9 that there was a perfectly blue sky, and Wolcott returned to the plantation with her cameras, where she took photographs of the main house, Lyle's cabin, and other structures on the property. Wolcott and Cammie never met each other, as Cammie did not return from Hot Springs until the day after Wolcott left the Cane River area.

By this time the Southern Renaissance that had blossomed during the 1920s and 1930s had evolved to a point, at least with regard to literature, that left Melrose behind. The greatest works of writers such as Ada Jack Carver and Lyle Saxon had been produced, and Cammie, now almost seventy years old, no longer was needed to patronize their careers. The pilgrims who came to Melrose in the 1940s used Cammie's library for research or just wanted to see and walk about the gardens. Still, the allure of the plantation and Miss Cammie remained. In May 1940 Lillian Trichel wrote a letter to Carrie Dormon, after returning from a Sunday afternoon visit to Melrose, saying she found "Cammie as nearly at leisure as she can ever be" and wished she could visit more often: "There is no one in the world to match her, and no other place with the flavor of Melrose . . . I need Miss Cammie and Melrose."[12]

Like Lyle, Carrie had become too busy to spend the same amount of time at Melrose as she had in the past. In December 1940 she had committed her time to a project that fulfilled a decades-old dream. In her biography, Fran Holman Johnson notes that Carrie had been concerned about the landscaping of Louisiana roadsides for years, traveling all over the state, publishing numerous articles, and giving countless lectures about roadside beautification. She had a number of specific ideas to improve conditions, and Prescott Foster of the Louisiana Department of Highways hired her to at last carry out her objectives. The job was created for Carrie, and the lack of precedent created several hurdles for her, but she reveled in the work. Cammie was ecstatic, telling Carrie, "At last someone will plant the hi-ways who has tree sense . . . we must have beauty summer and winter."[13] Anxious to give her own input about the project, Cammie urged Carrie in letter after letter to come to Melrose so they could talk.

Reference to Ada Jack Carver in the correspondence between Cammie, Carrie, and Lyle eventually becomes nonexistent, although there is the occasional mention of her through common friends. In 1934 she gave a speech to the Shreveport Branch of the National League of American Pen Women, and Mary Belle McKellar, who was present at the event, typed a four-page, gossipy letter to Cammie with all of the scintillating details. She began the letter with a directive that "this must be destroyed," which Cammie of course did not do; in fact, she noted at the top, "Too good to destroy."

Ada's speech veered wildly off topic, and she at times seemed confused and disoriented. When introduced, Ada "descended the stairs . . . and proceeded to take her seat in the front row," rather than take the podium as expected. Once Ada finally began her talk, which was to be on stream of consciousness in the short story, "she evaded the subject, talked at length about a scrap book she was making for her son" and all that was going into it.[14] Biographer Oliver Ford points out that these women's meetings of the period were almost farcical by nature and that "the devastating lack of real understanding and sympathy from members of her own social group" is heartbreaking in that Ada truly seemed to be suffering some sort of mental breakdown.[15] There is no way to know if Cammie reached out to Ada during this time or what she may have said, as all of Ada's papers, letters, and any potential stories she may have worked on were destroyed upon her death.

One of the last communications to Melrose from Ada was in 1938. Cammie Jr., then twenty-three years old, married Dr. Eugene Wenk in September of that year, and they moved into writer Kate Chopin's former home in Cloutierville. That Christmas, Ada sent a pair of candlesticks along with a note, saying: "I've been told it's a good omen when a bride receives candlesticks on her first Christmas and so I'm sending you a pair for the lovely old house. Perhaps we shall drop in one day during the holidays."[16] It was a charming gesture and typical of Ada's gracious nature and good manners.

In a December 6, 1941, letter, a friend in Shreveport wrote Cammie that Ada had made a most bizarre appearance in his furniture store, dressed in a gray muff and veil and with two eyebrows penciled on one side of her face.[17] The incident seemed to reflect Ada's increasingly fragile mental health. Donnis Taylor's 1991 dissertation on Ada Jack Carver portrays her last years as a tangled spiral into dementia, describing, for example, Ada sending her maid to nice shops for expensive dresses that she would then cut apart with scissors. She quit replacing burned-out lightbulbs in her house and used candles instead.[18] Eventually, family members became concerned for her safety and moved her into a nursing home in Minden.

Of the original inner circle at Melrose, Ada was perhaps the most emotionally vulnerable. Her sense of propriety and her artistic sensibilities, combined with the tragedies she had faced as a young woman, shaped her to view the world of civic clubs and women's groups as dull and tiresome but necessary to her existence and her perceived role. In 1925 she once told Carrie how she truly felt about the society scene: "Only God in Heaven (and maybe you) knows how I loathe and despise these things where a lot of women talk and jabber about nothing for hours. I always come back from them completely sapped of all strength of mind and body. I believe I give a piece of myself to every person I come in contact with, and when I'm alone once more there's nothing left."[19] Still, she was

very happy in her marriage and adjusted to her life in Minden, though Natchitoches never left her, and neither did Melrose. Cammie, Lyle, and Carrie were
among the few people with whom she could be herself and express her feelings
freely. In February 1946 Ada's mother died, and in March 1959, her husband
died. Carver herself wrote later that "February, March, and April mean to me a
series of hallowed but heart-breaking anniversaries."[20]

In the early years of the 1940s, Cammie and François were transcribing the various diaries they had collected, Alberta Kinsey came to Melrose to paint, and
Cammie's grandchildren were about the place. Joe Henry, who at this time lived
in Beaumont, Texas, allowed his son, Pat, to stay there during the summers, and
Joe was "mighty happy" because the child was thriving at Melrose.[21] Cammie
would often have him climbing up in the fig trees to shake the fruit loose so she
could make preserves. One day young Pat built a pair of stilts. François recounts
that both he and the houseman, Frank, tried them out, but Cammie refused, saying, "You all will be the death of me anyway, so there's no need for me to break
my neck on those things!"[22] At seventy years of age, Cammie had better sense
than to try to walk on stilts, but the anecdote reveals the feisty spirit and sense
of humor that seemed consistent throughout her life, whether she was talking
to servants, friends, or grandchildren.

Roosevelt's New Deal, particularly the Civilian Conservation Corps (CCC)
public work relief program, and the war were contributing factors in the death
of working plantations like Melrose. It had become very difficult to find and keep
workers, sometimes making it hard to get the cotton in from the fields or keep
the gin running. The men had either gone to war, gone to work for the government in the CCC, which paid better wages, or had enlisted in military service. In
1942 Lyle wrote Carrie: "All the War news is unbelievably bad. I hope bombs do
not destroy your pond and iris. I seldom get to Melrose any more."[23] Of course he
was teasing about the bombs at Briarwood, but it is true that his visits to Melrose
had become seldom, and the plantation was not what it used to be. Lyle loved
the idea of the plantation in general, the plantation system perhaps, and it no
longer existed, despite both his and Cammie's efforts to retain it both practically
and through vicarious means such as Lyle's literature. Cammie still wore the long

skirts and shirtwaist that had been her trademark since she was a young girl. She still maintained, as best she was able, the manners and mores of the Old South at Melrose, keeping as many servants as she could manage to help run the house and the grounds. Dark, black coffee was still delivered to the rooms of guests every morning before breakfast in the tradition of times past.

Lyle's work with the Works Progress Administration (WPA) had required his presence in Washington, D.C., and in February 1943 he wrote to Carrie that he was there to finally close the WPA office. With the guidebook done and his other editorial duties finished, he was ready to return to Louisiana for good. He visited Melrose briefly that summer and in typical fashion used his sharp wit to make fun of certain people he thought pompous or superficial, composing a letter to Cammie in which he pretended to be a prominent Natchitoches society woman they both knew: "Dear Mrs. Henry: What a pleasure to visit you in your quaint old plantation home! Join us tonight in a game of bridge or poker. All of your friends will be there . . . The refreshments will consist of the most delicious viands, including hot chockolate [sic] gin fizz, oatmeal soup, canned salmon and ice cream salad and a delicious dish of tripe a la mode. Pink mints and chocolate covered pecans will add a piquant touch. Do not fail to join us, Cammie dear, for we are all dying to see you." At the bottom of the letter, Cammie penciled in a date and wrote, "A perfect Lyle Saxon mss."[24]

Lyle had not truly lived at Melrose for almost eight years, making only occasional weekend visits. François Mignon was occupying Lyle's cabin most of the time, having abandoned his "maisonette" for the comfort of Yucca's books, fireplace, and accommodations. When Lyle visited Melrose in July 1943, he wrote to a friend that he found the plantation run-down, the gardens neglected, and the cabin "dirty and full of spiders."[25] In a letter to Carrie, he said Melrose now "seems a sad place, somehow." The grounds were overgrown, and Cammie, then seventy-two, was fighting chronic malaria, which kept her in bed a good bit of the time with crushing headaches.

Though she was slowing down physically, in her heart Cammie still had one foot on the road. In March 1943 she wrote to Carrie: "Let's steal one day to play—trip do me good—I'll pay for gas and oil—please let's do this—you and I never play & the day is far spent—our journey not so much longer—we've got to do a few things before we fold up."[26] She also retained her curiosity, tapping into Carrie's gardening expertise by inquiring: "Please tell me before either of us are any older—why magnolia leaves are not good as mulch? I have got to know!"[27]

In October 1943 New Orleans author Harnett Kane came to Melrose to conduct research on old plantation homes. Kane had read the Alice Premble White
thesis on Joseph and Lavinia Erwin, and he wanted to include the Andrew Jackson duel with Charles Dickinson as well as the story of Shady Grove in his new
book. Cammie agreed to talk about her ancestor with him, so Kane made several
visits through the end of 1943. After that, he and Cammie corresponded back
and forth, she answering his questions. In July 1945 he wrote to her about his
forthcoming book, *Plantation Parade:* "Morrow reports that they think the Melrose chapter the best; it is probably the longest and fullest in the book, covering
about 22 pages in my type. The Erwin chapter covers about 15 pages dealing with
the first Erwin place, Dickinson's Live Oaks, and the St. Louis plantation which
went up near the site of the first Erwin place . . . there are about 40 pictures of
plantation houses . . . Of Melrose I managed to get in the big house and also the
African House."[28]

Everyone at Melrose waited with anticipation for Kane's book to hit. By October, Cammie still had not received a copy, but she was certainly aware of what it
said and was not pleased. François wrote to Carrie: "In his chapter on Melrose he
got every historical fact wrong, and came pretty close to libel in the impression
he gave of the Madam, wherein he speaks of her, according to one tale, as feeding
one Writer—Lyle of course, through the transom, after locking him in his room."[29]

Cammie was furious at Kane's representation of her. In a letter to her friend
Irene Wagner, in April 1946, Cammie said: "That book of Harnett Kane made us
all [mad]—all so wrong. Dan declares Kane shall *never* visit Melrose again . . .
he does not take the trouble to be accurate—my friends in St. Francisville are
all furious . . . Kane's picture of me was a *caricature* not a portrait . . . every place
he wrote about was wrong . . . well it's just lies—*money* is all that counts. J.H.
only read the first part of Melrose and said all of Isle Brevelle is inaccurate."[30]
Cammie felt as though Kane should have let her read the manuscript first, at
least as far as it applied to her and her family. Kane never offered, and apparently
she never asked. She attempted to talk about it with Lyle, but he told her that of
course there was nothing he could do about it and that if he tried to address her
issues publicly, it would simply make him appear jealous of Kane's book. So, in
the end, there was nothing to be done. "I'm so sorry he gave me that publicity,"
she told Wagner.

The title of the chapter about Cammie is "Chatelaine in Shirtwaists," and
Kane described her as "a Louisiana phenomenon, like a bayou, a Cajun dance, or

a crawfish digging under a levee." He wrote that Cammie had grown up among books, had gone to the Normal School at Natchitoches, and was a "female with opinions."[31] None of that is particularly offensive; what seemed to have riled Cammie most was the passage about her locking "a writer" in his room, and when "the harried fellow" passed his written pages under the door to her, had they "not been up to snuff, Aunt Cammie would have entered with a cowhide whip left from slavery days and threatened to use it."[32] It is difficult to imagine Cammie ever taking a rawhide whip to Lyle, the writer to whom Kane must have been referring. Irene Wagner confirmed later that Cammie regarded the section about her to be a caricature and that "the spirit of the thing," especially the chapter on the Erwin family, was offensive for portraying her ancestors as uncultured and with lofty airs.[33] Both chapters have been quoted often since Kane's book was released, and now, too often, the facts cannot be discerned from the truth.

By 1945 the American photographer Frances Benjamin Johnston was living at 1132 Bourbon Street in New Orleans, in close proximity to Lyle Saxon, Alberta Kinsey, and others in Cammie's circle of friends. Johnston was a well-known and celebrated portrait and news photographer who at that time was working on photographing early American buildings and gardens. She received several grants from the Carnegie Foundation, enabling her to travel the South and photograph architecture, including remains of old plantations. Before Uncle Sam Plantation in St. James Parish was demolished in 1940, Johnston was able to photograph it, thus preserving on film a plantation Cammie had documented in her scrapbooks through her own photographs and clippings.

Kinsey was likely the person who introduced Johnston and Cammie, who shared interests in gardening as well as preservation. In the summer of 1945, Johnston visited Melrose for at least a week. In a 1946 letter to Cammie, she wrote of her Bourbon Street house, which she was converting from a single dwelling to three apartments, and of her tribulations in renovating her courtyard into a garden. During her stay at Melrose, Johnston and one of Cammie's employees had walked the gardens, tying red flags on nearly twenty trees, shrubs, and bulbs that Johnston wanted transplants or cuttings from for her own garden. "I would like any plant material you feel might be spared without causing trouble or inconvenience—or leaving holes in your own garden," Johnston wrote.[34]

While at Melrose in 1945, Johnston took photographs of Cammie sitting in the gardens, a large shawl draped over her shoulders. One photo also includes François Mignon standing erect behind Cammie, his hand placed on the back of the bench where she sits, as they both gaze at the house. In his journal François identified Johnston as the photographer of the series of images, remarking that while "the picture of the Madam is alright . . . it might suggest 'Whistler's Mother.'" He goes on to write, "For myself, I never saw the Madam in her garden relaxed as in this particular picture . . . she was always as busy as a bee, flying from weed to wee[d], or from flower to flower, but never in the reposed attitude of the picture."[35] François grumbled that after Cammie's death, when Cammie Jr. asked Johnston for a copy of the picture of her mother, the photographer offered to sell her a glossy print for five dollars. "I think that is rather hilarious," he said, given that she "was a guest in this house for days on end." Mignon added, "No matter how much or how little she asks for the prints of the pictures she took of her hostess while a guest in this house, it doesn't matter in the slightest to me, for by her price, disregarding past courtesies and hospitality, she does more to establish her own portrait than anything her old camera ever recorded of anyone else."[36] Nevertheless, the photograph of Cammie remains an iconic image of her as well as a valuable contribution to Johnston's body of work.

Only three days after Cammie's irate letter to Irene Wagner regarding Harnett Kane's book—on April 9, 1946—Lyle Saxon died in New Orleans at age fifty-four. According to biographer James Thomas, he underwent surgery in April but never recovered, dying of bladder cancer not quite two weeks before Easter. It was Easter that first brought Lyle to Melrose, and it was on Easter cards that Cammie began receiving condolences from friends on his death.

One of his last visits to Melrose was the previous August, when he stayed two nights.[37] That October, in New Orleans, he joked about both his and Cammie's maladies, writing he was sending a cane to her via J.H., who had come to visit him. "Do use it," he urged, "keep it by your sofa, and you'll find out how much easier it makes it to get around." He also expressed nostalgia for past times at Melrose: "It makes me think how long it has been since I've really lived at the cabin . . . Lord, God, it seems like a lifetime. Too much has happened to us." He closed with his usual good humor, saying: "I'm counting on you to get well again so that you and I can make a comeback that will bring confusion to our enemies. I love you very much, Ma'am."[38]

Lyle felt well enough during Carnival 1946 to write his usual bit about it for the *New Orleans Times-Picayune* and to do a live radio broadcast from a balcony during the Rex parade.[39] After Mardi Gras, in February, he wrote to Cammie that he was coming home to Melrose to rest, and she must have overseen the usual preparations, such as having his cabin cleaned and placing fresh flowers in his rooms. Saxon's friend Joe Gilmore phoned the plantation in March to tell Cammie they were coming, but by the time they reached Baton Rouge, Lyle began hemorrhaging in his stomach, and his friends turned the car around and drove him straight to the Baptist Hospital in New Orleans.

Cammie apparently was not fully aware of how seriously ill Lyle was because in her letter to Irene Wagner, she wrote, "Lyle took sick to drink nearly killed himself—is in a hospital right now—he had his nurse write me—all she said— Lyle was doing as well as could be expected—which tells me nothing." Cammie did receive short warning on the afternoon of April 9, in a telegram from Alberta Kinsey stating that Lyle's condition was critical and that his recovery was doubtful. And later that afternoon, family friend Charles Mazurette phoned the plantation to report that "Lyle's death was expected hourly."[40] On April 14 Lyle's friend and fellow author Robert Tallant wrote to Cammie of his last days: "I want you to know that he passed very quietly, very peacefully, and without pain. During the last day his breathing was harsh and a little difficult, but he was unconscious the entire time. For the last few hours his breathing became softer. And then it simply stopped. That was all there was to it. I was in the room."[41]

That was all there was to it, indeed. Lyle's death was the beginning of the end of Melrose as it had been in its halcyon days of the writers and artists. Kinsey wrote Cammie on April 13, and it appears from her words that Cammie was taking this loss in her usual stoic fashion: "I am glad you are such a philosopher and can reason things out. I haven't been able to do so yet and feel as if the props have been taken out from under me. Of course will get over it."[42] Cammie likely reasoned that if Lyle could not be himself, if he had to live in pain and as an invalid, then it was better that he go. He would not want to live that way or to be a burden.

Mazurette, who had retired to a place on Little River just six miles from Melrose, wrote to Cammie from Baton Rouge: "His body lay in state where all his neighbors of long ago as well as many friends and acquaintances could come to pay tribute. One of Baton Rouge's most talented and celebrated sons had at last come home to rest."[43] While "home" for Lyle was technically Baton Rouge, it was also New Orleans. Services for "Mr. New Orleans" were held in both cities: there was a memorial event in New Orleans at the funeral home, and in Baton Rouge a funeral was held at St. James Episcopal Church, with burial in Magnolia Cemetery. Because of her health, Cammie did not go to either service, but Cammie Jr. and her husband, who served as one of Lyle's pallbearers, attended the Baton Rouge funeral.

Cammie learned that Lyle's aunts were planning to leave his letters and manuscripts to the Louisiana Library Commission, and on April 20 she contacted the commission in Baton Rouge. "I have much material in books, records, magazines, and clippings that would add materially to such a project and I will be glad to contribute whenever the Louisiana Library Commission can provide adequate space for such a collection."[44] Ultimately, Lyle's papers were deposited at Tulane University. Cammie went to her scrapbooks and wrote: "Lyle died Tuesday, April 9, 1946[,] at 9:20 p.m., at the Baptist Hospital N.O.—died of pneumonia following unsuccessful operation for cancer—burial in Baton Rouge in Magnolia Cemetery. Left no will—& his aunt Maude Chambers is legal heir, gave his personal effects to Tulane."[45] Despite her unemotional tone, one must assume Cammie felt pain in writing those words. But in her mind, it must all be recorded in the scrapbooks.

N ow seventy-five years old, Cammie took longer to rebound from illness than she had in the past. In September 1945 François Mignon recorded in his journal that Payne Henry had been to Hot Springs, where Cammie had gone on one of her regular trips to "take the baths." In speaking with Cammie's doctor there, Payne learned that she had been prescribed "liberal doses of codine [sic]" by one of her doctors in Louisiana. Mignon suggests it had been prescribed by Dr. Wenk, her son-in-law, and it is true that Wenk was often on hand to attend to Cammie's increasing medical needs. However, she also saw the country doctors who had taken care of her for years, and there is no way to know for certain who prescribed the medication. As an opiate, codeine is addictive, causes drowsiness, and can cause stomach problems. Use of this drug could have been a contributing factor to Cammie's loss of vigor and growing reluctance to leave Melrose unless absolutely necessary.

François recorded on January 13, 1946, that Cammie had perhaps suffered a light stroke in previous months, noting that upon checking on her in the afternoon, he found she had slid out of her bed and "she sat on the floor, being unable to get up." When she walked, she dragged one foot slightly, which had prompted Lyle to send her the new cane the previous fall. Mignon's entries over the next few months indicate that Cammie improved slowly, sometimes coming downstairs for meals and sometimes not. She continued her correspondence, sometimes writing her own letters and sometimes dictating as François wrote them. She also continued pasting items in scrapbooks, but clearly her overall health was in decline.

Between 1946 and 1948, Cammie spent much of her time in her room. She read books, wrote letters, and listened to a small radio she kept by her bed. She seldom came downstairs for dinner, taking meals in her room. In early 1948 she suffered a stroke that furthered her confinement upstairs. J.H. acquired a wheelchair for her, but Cammie refused to sit in it. According to François, she told J.H., "Well, you needn't get it because I'm not dreaming of dying and I won't ever sit in the thing."[46] In April, Cammie Jr. wrote to J.H., requesting consent for her mother to visit her in Shreveport, where she and her husband then lived. J.H. vehemently vetoed the trip, feeling that the travel and the company of Cammie Jr.'s young family would be too stressful.

In early May, Alberta Kinsey visited the plantation and stayed for two weeks. Cammie enjoyed the company, and in the evenings the little group would load into J.H.'s Cadillac and take a drive around the Cane River. Cammie failed to recover her strength or balance, however, and grew frailer as the year progressed. François wrote about her fatal fall on November 17, 1948, in his book, *Plantation Memo*: "Mrs. Henry had not been well, and each night I made a round by her wing of Melrose every other hour. It was after four in the morning when I discovered that she was not in her accustomed place, and on further examination I found that she had apparently toppled from the hassock, striking her head against the sharp edge of a doorstop. She never regained consciousness."[47] The doctor was called, and the ambulance came, but there was little chance that Cammie would survive the injury. The official cause of death listed on the death certificate is skull fracture and cerebral hemorrhage.[48]

News of Cammie's death quickly spread across the state and the South, to the shock and bereavement of the many beneficiaries of her patronage. It was reported in major newspapers throughout the South. The *Times-Picayune* referred to her as "mistress of famed Melrose plantation and Louisiana patroness of the arts" and noted her extensive library as well as the luminaries who had studied and worked there. As messages and telegrams began to pour in, the local Western Union office had to bring in extra clerks. The florists could not keep up with the orders. Every Henry grave in town was adorned with flowers. Finally, Stephen Henry decided to convert the unfilled orders to a standing order for fresh flowers for the next year in the research center in Natchitoches established in Cammie's name, rather than having them "wilting away in the cemetery."[49]

The day of Cammie's funeral brought a deluge of rain, which ended a drought that had lasted for some five months around Cane River. The Henry children gathered at the plantation, and the morning of her funeral was a flurry of preparation and emotion. François Mignon poignantly recorded every detail of the day in his journal, from arrangements of magnolia leaves "bursting forth" from the fireplaces to the positioning of the chairs in the library, where Cammie's body was laid out. Both J.H. and Stephen—whom François called "The General," reflective of his military rank—had spoken to François the night before and asked him to stay on indefinitely at Melrose after the funeral. Both of them offered him a suit to wear to the service. François and Stephen shared a moment outside in the rain, shedding tears while remembering the great lady they prepared to honor.[50]

According to François, before the service, he and Joe also shared a word. He quoted Joe as having said, "François, if I was ever reduced to ten cents, that would still make two nickels and one of them would be yours for the asking." Then they, too, shed tears before François went to Yucca House to change clothes for the funeral.[51] Cammie Jr. and her family arrived, and the wives of the Henry boys were all present, mingling with guests and reminiscing.

The downstairs library and adjoining dining room had been cleared of the usual furniture, and Cammie's casket lay at the end of the library, covered by a blanket of American Beauty roses. Cammie Jr. sat somberly, quietly, and alone in the dining room throughout the morning. Robina Denholm came, as did Carrie Dormon and many dignitaries and professors from Northwestern State College (formerly Louisiana Normal College) in Natchitoches. Prior to the two o'clock service, François wrote, "The negro and mulatto friends of the Madam were equally ushered into the big quiet library where she lay in state." As the service began, "the heavens opened up, and all the pent up water of half a year came cascading down."[52] Chairs had been arranged in long rows the length of the library, with a center aisle; François sat between Joe and Payne Henry. He reported that the Presbyterian rite was somewhat impersonal, which was a sort of blessing, as it kept overflowing sentiment at bay. Following the funeral, everyone waded through the sodden gardens to the cars waiting to take them to the American Cemetery in Natchitoches.

François rode with J.H. and Celeste; Stephen and his wife, Mamie Gray, followed behind them, bringing along two members of the Melrose kitchen staff. Following them was "the rest of the family, and then the friends and acquaintances, including both colors, which was precisely as the Madam would have had it," François wrote. The rain tapered off as they reached town, but once at the cemetery, it again began to pour. This time François found himself seated next to Cammie Jr. under the canopy, which was doing little to protect everyone from the rain. Thirty years after her husband's death, Cammie was laid to rest beside him in the Henry family plot.

At the conclusion of the graveside service, Cammie Jr. sat alone by the casket as everyone began to make their way back to the cars. One of her brothers spoke gently to her, urging her to come along, and finally young Cammie stood and prepared to depart, saying goodbye to J.H. and giving a solemn, formal nod to her brother Stephen as she and her husband departed for their own home. In many of his diary entries, François indicated some tension between Cammie Jr. and her brothers. She was never very close to Stephen or J.H., due at least in part to the twenty-year age gap between them. François's reportage of Cammie Jr.'s behavior often hints at a spoiled, willful young woman who brought stress and turmoil whenever she visited Melrose. He referred to her Sunday visits as "Wenk-ends" and suggested that these visits disrupted the day-to-day routine of the plantation. While François seemed to remember his place as a non–family member, he was, at least in his writing, protective of Cammie and somewhat judgmental of her daughter. His account of Cammie's funeral is likely accurate, but it would also be fair to consider that his apparent disapproval of Cammie Jr. might have colored his perspective.

After the service, small talk on the way back to Melrose included observations about the weather and the large volume of Melrose pecans that certainly must have been shaken from the trees by the winds and more than likely ended up in the Cane River, "floating away in the ditches . . . halfway to Alexandria" and on their way to the Red River. At the big house, the dining room furniture had been returned to its place, and the smell of fried chicken, scrambled eggs, and coffee awaited the grieving friends and family.

The Henry children decided to keep Cammie's bedroom just the way she left it each night when she went to bed, to preserve it with all of the effects of her personal sphere, including her books, woven fabrics, and the gold brooch that had belonged to her grandmother Erwin and that Cammie was seldom seen without. Presented by Isaac Erwin to his bride, Carmelite Picou, upon their marriage, it had passed to Leudivine and then to Cammie. It was a relic and a symbol of days gone by.

Epilogue

A Landmark Ought to Be Preserved

*I enjoyed every minute of last summer . . . those long evenings when
we sat on the front gallery in the moonlight and talked.*

—LYLE SAXON, letter to Cammie Henry, n.d.

I f Cammie Henry could have written her will in the same telegraphic way she
wrote her letters, scrawled on the back of an envelope or on some scratch paper, she probably would have. As it is, her handwritten will is only three sen-
tences. In the first, she affirms that this is indeed her last will and is in her own
handwriting. In the second, she requests that all of her estate "go to my children
to be equally divided among them." And in the last sentence, she appoints J.H.
and Stephen as executors. There are no special bequests, no instructions in the
will for her library, nothing more to say. The will was drawn on February 9, 1922,
twenty-six years before her death. It is puzzling that in all those years, she never
updated her will or made more specific provisions for her vast collection of materials, antiques, furniture, and other personal property. Perhaps she always intended to but never managed to do it; perhaps she felt that J.H. and Stephen
would intuit her wishes and carry them out. Perhaps she just never saw the need.

After Cammie's death, François Mignon stayed on at Melrose for almost
twenty-two more years. He continued his correspondence with Caroline Dormon and with Robina Denholm, and although Cammie was no longer there,
tourists kept coming. François called them "passing pilgrims," welcoming them
and showing them the grounds. His value as curator and preservationist of Melrose is important. If J.H. kept the business end of the plantation going, François's

continued presence kept the tourist end of the plantation alive. He talked to visitors about its history and wrote about it in his newspaper column, "Cane River Memo," later called "Plantation Memo," which drew people to Melrose to see and experience the Land of the Lotus for themselves.

J.H. continued to successfully run the plantation until his death in 1969. Then the family took a vote and decided that since they had a buyer for the property who was not interested in the contents and the family was too large for everyone to get their favorite pieces, the contents must be sold at auction. It seemed the only fair way. Family members would have to compete with collectors, preservationists, and the general public for Cammie's things. Cammie had been gone for just over twenty years at that point, but inspection of the extensive number of items on the auction list shows that if any of her possessions had been removed from the plantation in the meantime, the amount was negligible.

The auction was held June 6 and 7, 1970, at Melrose. François did not attend—he could not bear to. Carrie would come nowhere near it. Over two days, masses of people swarmed the grounds of Melrose, gawking, waving bid paddles, and carrying off everything on-site. Literally everything was sold, from crockery to quilts, rocking chairs, Cammie's looms, the treasured McAlpin stencil, cups, saucers, lampshades, right down to two pitchforks and a box of buttons. Some of the items were designated as having belonged to Lyle Saxon. "Lyle Saxon's walnut armoire," for example, sold for one hundred dollars; and "Lyle Saxon's andirons" sold for ninety dollars. The famed portrait of Grandpère Augustin sold for two thousand dollars and now hangs in the church that he founded on Isle Brevelle. Several Clementine Hunter paintings, quilts, and dolls were sold, for a fraction of what they would fetch today. Some items were purchased by members of the local historical society, family members, and local arts patrons, with the hope of eventually returning them to Melrose, and in fact, a great many of Miss Cammie's things are back in residence.

It had been agreed among the children that their mother's scrapbooks and library would be donated to Northwestern State College (today University) of Louisiana. When the collection was appraised in 1970, noted historian and archivist Winston De Ville judged the scrapbooks alone to be worth $30,000, commenting that the materials in the books "are rarely found so well-gathered and maintained."[1] The Melrose Collection at the Cammie Garrett Henry Research Center at Northwestern State University remains a fascinating collection not just for Cammie's own scrapbooks and correspondence but also for the historic documents, early manuscripts, and personal diaries that she was able to assemble. It is a valuable resource for all who still come to Miss Cammie's library to do research and turn the pages of history. She would be pleased.

Southdown Land Company, an agribusiness, was the purchaser of the Henry family's land. Recognizing its historical importance, Southdown donated over six and a half acres of the property, including the house and outbuildings, to the Association for the Preservation of Historic Natchitoches, which then worked to have Melrose declared a National Historic Landmark in 1974.

After Cammie's death, Carrie continued her work in forestry and preservation and published several more books and many articles for journals and magazines. She died in Shreveport in 1971 and is buried near Briarwood in Natchitoches Parish. In the summer of 1949, Ada Jack Carver published one more short story, "For Suellen with Love," in the *Centenary Review,* a literary magazine produced by Centenary College in Shreveport.[2] Carver's biographer Oliver Ford suggests this story, or at least the publication of it, is a sort of tribute to Cammie, "who had nurtured her, supplied her with stories and facts, and been her only confidant."[3] It was as if Ada had dug deep within herself to publish just once more for Miss Cammie. There were no more published stories from her after that. She died in 1972 in Minden, Louisiana. Alberta Kinsey visited Melrose several times after Cammie's death to paint. In February 1949 she wrote to François that she was considering a trip to Melrose, but François sensed that Kinsey was a little "frightened in the country when night settles down," and her visits soon ceased.[4] Kinsey died in 1952 in New Orleans. François left Melrose in 1970, when the family sold the property, to live in a place on the river closer to Natchitoches. He died in 1980 and is buried in St. Augustine Catholic Church Cemetery on Isle Brevelle next to Clementine Hunter, who died in 1988.

Melrose draws about 100,000 visitors each year.[5] The "Myth of Melrose" female trifecta is of course Marie Thérèse Coincoin, Carmelite Garrett Henry, and Clementine Hunter. The power of these three women has proven so alluring for the local tourist industry that now it is sometimes difficult to extricate truth from fiction. But Melrose and Cammie Henry are indissolubly linked.

Melrose's continued existence today is due to her foresight and hard work to preserve the historic buildings and to nurture and support many artists, writers, and historians of her time. Her lifestyle reflected the gracious, genteel, and mannered ways of the Old South that she idealized even while she put her own wry, fun spirit into everything that she did. Her style of dress and her continuation of old plantation traditions pointed to the past, but her work in preservation looked to the future. An enigma, she lived life to the fullest up until the very end, following the philosophy she once shared with Carrie Dormon: "Life is too short to be idle."

Today the trim, manicured gardens of Melrose look very different from the riotous wild growth that covered the grounds when Cammie lived there, but the mystique of Melrose has remained. After his first visit on Easter Sunday 1923, Lyle Saxon wrote in the *New Orleans Times-Picayune* that when he died and went to heaven, he would bypass all "of the celestial grandeur" and ask only "to spend the rest of time in visiting the places on earth that pleased me so much when I was alive." He said that one of the first places to which he would return "will be Melrose."[6] It is easy to believe that Lyle's spirit has done just that when one sits on the front gallery of his cabin, which is almost hidden behind beds of towering wild ginger and giant banana plants with unusually large purple summer blooms.

The cabin is different, though—not really Lyle's cabin now. Brightly white-washed, with fresh paint on the shutters and a brick gallery at ground level both front and back, it is picture perfect. All the doors and windows are thrown open for tourists to peer over velvet ropes and view the inside, the same kind of tourists who so often distracted Lyle from his writing. There is no life-sized portrait of Grandpère Augustine Metoyer hanging on the wall. The shelves are empty of Lyle's library of eclectic books and his hidden jug of liquor. Cammie's handmade rugs no longer cover the floor, and Lyle's custom, oversized bed is gone. There is no longer a crackling, welcoming blaze in the fireplace.

Still, the mistress of Melrose would likely approve of the fact that her house remains open to visitors. She would offer them a cup of strong coffee and invite them to step back in time, walk through the gardens, page through a few scrapbooks, and talk for a spell. At dawn the mist still lifts off the Cane River, the morning sun kisses Miss Cammie's garden, and the plantation comes to life.

NOTES

ABBREVIATIONS USED IN NOTES

CDC	Caroline Dormon Collection
CGHRC	Cammie Garrett Henry Research Center, Northwestern State University of Louisiana
FM	François Mignon Collection (CGHRC)
FMCHC	François Mignon Collection, Southern Historical Collection, Wilson Special Collections Library
FWPC	Federal Writers Project Collection
MC	Melrose Collection
SB	Scrapbook
SGHC	General Stephen G. Henry Collection

PREFACE

1. Ada Jack Carver to Cammie Henry, September 4, 1924, SB 217, 9, MC, CGHRC. The reference to the mythological Lotus Eaters who left Odysseus and his sailors blissfully happy with no interest in returning to their ships could refer to Ada's life in Minden, which often included a busy social calendar that kept her from her writing.

2. Lyle Saxon, "Easter Sunday at Aunt Cammie's," *New Orleans Times-Picayune Magazine*, May 20, 1923.

3. Kane, *Plantation Parade*, 271.

1. FINDING HER WAY, 1871–1920

Epigraph: Cammie Henry, letter to her son Stephen, March 30, 1911, box 1, folder 6, SGHC, CGHRC.

1. Cammie Henry handwritten marginal notation, SB 18, 17, MC, CGHRC.

2. *Dictionary of Louisiana Biography,* 1:287.

3. Cammie Henry handwritten marginal notation, SB 18, 15, MC, CGHRC.

4. White, "Plantation Experience," 348.

5. Kane, *Plantation Parade,* 207. The book includes chapters on both Cammie Henry and her great-grandfather Joseph Erwin. It should be noted that Cammie was very displeased with the chapter about her as she felt Kane had reduced her to a caricature.

6. Erwin family history typescript (unattributed, folder 1211, MC) correlates with Kane's version in *Plantation Parade.*

7. White, "Plantation Experience," 377 and 399.

8. Wenk genealogical papers, folder 2-J-5, Wenk Collection, CGHRC.

9. Cammie often wrote marginal notes in her scrapbooks; she records this information in SB 9, 72, alongside Barrow's *Progressive Farmer* article.

10. Cammie Henry marginal notation, SB 9, 72, alongside a 1911 article about the death of Judge Charles E. Fenner, who organized Fenner's Louisiana Battery at Jackson, Miss., Cammie noted, "My father S.R. Garrett served in Fenner's battery." Specific information regarding Captain Garrett's service can be found in Andrew W. Booth's *Index to Louisiana Confederate Soldiers,* published on U.S. Gen Web Archives, accessed June 10, 2014.

11. Cammie Henry described the location of Scattery Plantation, SB 18, 35, MC, CGHRC.

12. "Biographical/Historical Note," Miles Taylor and Family Papers, Hill Memorial Library, Louisiana State University Libraries, Baton Rouge, La. accessed May 5, 2013, http://www.lib.lsu.edu /sites/default/files/sc/findaid/1378.pdf.

13. Stuart, *More than Petticoats,* 103. Stuart says Cammie "attended private schools in Donaldsonville."

14. The banknotes are in MC, CGHRC, folder 1097, along with a letter from Charles Garrett of North Carolina to Leudivine dated November 1884. The banknotes are dated January and March 1885. Additionally, there is a loan from James Comfort to Charles Garrett for Leudivine.

15. "Alumni Number," *Normal Quarterly* (Louisiana State Normal College) 10, no. 4 (October 1921), accessed May 15, 2014, https://archive.org/stream/alumninumber1921nort/alumninumber-1921nort_djvu.txt. In SB 19, 82, Cammie affixed a postcard of the school and wrote: "Guion Academy, Thibodaux LA where I was principal—& taught 2 yrs—What memories it brings!" This note is dated 1907.

16. *Weekly Thibodaux Sentinel and Journal of the 8th Senatorial District,* July 16, 1892, *Chronicling America: Historic American Newspapers,* Library of Congress online, accessed February 25, 2014, http://chroniclingamerica.loc.gov/lccn/sn88064492/1892–07–16/ed-2/seq-3/. Article notes that Cammie Garrett, principal of Guion Academy, had returned from Ruston, La., where she took part in a superintendents' convention.

17. Ora G. Williams, "Cammie G. Henry," 20.

18. Wenk genealogical papers, Eugene Lloyd Wenk Collection, CGHRC.

19. *Biographical and Historical Memoirs of Northwest Louisiana,* 1890.

20. "Mr. Joseph Henry," obituary, *Natchitoches Enterprise,* January 26, 1899, *Chronicling America:*

Historic American Newspapers, Library of Congress online, accessed March 12, 2016, http://chroniclingamerica.loc.gov/lccn/sn88064317/1899-01-26/ed-1/seq-3/.

21. Quoted in François Mignon, journal entry, March 18, 1940, folder 262, scan 87, FMCHC, accessed April 12, 2015, http://dc.lib.unc.edu/cdm/singleitem/collection/03ddd/id/216613.

22. Biographical and Historical Memoirs of Northwest Louisiana, 1890; also, in SB 9, 73, Cammie's handwritten note: "Your father Jno. H. Henry began as a young man to work for R.M. in N.O., was with the firm 4 years."

23. Cammie and John's wedding announcement appeared in the New Orleans Times-Picayune, January 12, 1894. Ora G. Williams, in "Cammie G. Henry," also notes the wedding date.

24. Francis Marie Garrett, "Erwin Family History." This is a typed history of the Erwin family, dated 1972, in Cammie's collection, which obviously Cammie would never have seen.

25. Gary B. Mills, Forgotten People, 19–28.

26. Ibid., 31–36, 66.

27. Elizabeth Shown Mills, "Demythicizing History," 403.

28. Cammie Henry, letter to "My Dear Dear Friend," October 7, probably 1937. folder 208, MC, CGHRC.

29. Gary B. Mills, Forgotten People, 174.

30. Ibid., 81.

31. Shiver and Whitehead, Clementine Hunter, 40.

32. The condition of the main house when Cammie arrived has been described in Stuart, More than Petticoats, 104; and also Diane M. Moore, "Mistress of Melrose," 7. François Mignon also discussed the move from Derry to Melrose and the work that awaited Cammie upon her arrival at Melrose. Folder 260, scan 10, FMCHC.

33. Gary B. Mills, Forgotten People, 159.

34. J. Frazer Smith, White Pillars, 167–68.

35. Cammie Henry, letter to Stephen Henry, April 5, 1911, box 1, folder 6, SGHC, CGHRC.

36. Mignon Collection, folder 260, scan 10, FMCHC.

37. Cammie Henry's handwritten notes in Payne's baby book, folder 1193, MC, CGHRC.

38. Details of John Henry's corn prize are throughout Scrapbook 79. His brother Stephen won the contest the year previous and is detailed in SB 190.

39. Letter to Cammie Henry from "Your loving cousin, S.E.S.," 1917, folder 1278, MC, CGHRC.

40. Joe Henry, letter to Cammie Henry, January 28, 1918, folder 1482, MC, CGHRC.

41. John Hampton Henry's cause of death is listed as "cerebral hemorrhage" on his death certificate.

42. An article in the local newspaper reported John Henry's death. Natchitoches Enterprise, March 7, 1918, Chronicling America: Historic American Newspapers, Library of Congress online, accessed March 3, 2014, http://chroniclingamerica.loc.gov/lccn/sn88064317/1918-03-07/ed-1/seq-2/.

43. Isaac Erwin Henry to Stephen Henry, March 17, 1918, folder 452, MC, CGHRC.

44. John Hampton Henry's will, folder 1101, MC, CGHRC.

45. The findings of the Eleventh Judicial District Court of Appeals, Gorum v. Henry, no. 20243, can be found in Louisiana Reports, vol. 138 (St. Paul: West Publishing Co., 1916), 596, Google Books, retrieved April 4, 2016. John Henry was cleared of all charges brought against him by J. M. Gorum.

46. Stephen Henry to Cammie G. Henry, folder 400, MC, CGHRC.

47. Barry, *Great Influenza,* 92.

48. Ibid., 4.

49. Joseph M. Henry, letter to Cammie G. Henry, October 16, 1918, folder 1482, MC, CGHRC.

50. Isaac Erwin Henry, letter to Cammie G. Henry and his grandmother Leudivine, both dated November 3, 1918, folder 1478, MC, CGHRC.

51. SB 189, 17, MC, CGHRC.

52. Erwin Henry's obituary, written by Dr. Dunn, SB 189, 29, MC, CGHRC.

53. Stephen N. Garrett, letter to Cammie G. Henry, November 18, 1918, SB 189, 37, MC, CGHRC.

54. C. I. Thomson, Chamberlain Hunt Academy president, letter to Cammie G. Henry, November 18, 1918, folder 1479, MC, CGHRC.

55. C. I. Thomson, letter to Cammie G. Henry, December 10, 1918, folder 1479, MC, CGHRC.

56. Cammie Henry, letter to Erwin Henry, n.d., folder 199, MC, CGHRC.

57. Edward J. Young, letter to Cammie G. Henry, October 20, 1919, folder 1480, MC, CGHRC.

58. Cammie Henry, notation in scrapbook for son Bobbie Henry, SB 194, 9, MC, CGHRC.

59. "The Bristol-Nelson School: For the Care and Training of Sub-Normal and Backward Children," folder 295, MC, CGHRC.

60. SB 194, 43, MC, CGHRC.

61. Stephen G. Henry, letter to Cammie G. Henry, October 13, 1919, SB 194, 19, MC, CGHRC.

62. Handwritten note by Cammie G. Henry, SB 194, 47, MC, CGHRC.

63. Ibid.

64. Dan Henry, letter to Cammie G. Henry, January 1, 1921, folder 1460, MC, CGHRC.

65. Stephen G. Henry, letter to Cammie G. Henry, January 21, 1921, folder 405, MC, CGHRC.

66. Stephen G. Henry, letter to Cammie G. Henry, February 6, 1921, folder 405, MC, CGHRC.

67. Leudivine Garrett, letter to Dan Henry, September 28, 1921, SB 186, 7, MC, CGHRC.

68. John H. Henry Jr., letter to Dan Henry, September 1921, SB 186, 11, MC, CGHRC.

69. Handwritten marginal note by Cammie Henry, SB 186, 19, alongside a letter from J. H. Henry Jr. to Dan Henry, December 14, 1921, MC, CGHRC.

70. Cammie Henry, letter to Stephen G. Henry, March 4, 1924, folder 404, MC, CGHRC. Cammie never sent this letter, noting at the bottom of it, "Never sent—no use," yet she kept it.

71. Mamie Gray Henry, letter to Cammie Henry, June 30, 1924, folder 412, MC, CGHRC.

72. Stephen G. Henry, letter to Cammie Henry, July 1, 1924, folder 412, MC, CGHRC.

73. Cammie Henry, letter to Caroline Dormon, September 1924, folder 590, CDC, CGHRC.

74. Lyle Saxon, letter to Caroline Dormon, June 27, 1925, folder 555, CDC, CGHRC.

75. Jenny Smith, "Examination," 28.

76. Cammie Henry, notation in Payne Henry's scrapbook, SB 202, 15, MC, CGHRC.

2. CONGENIAL SOULS COME TOGETHER, 1920–1924

Epigraph: Mary Belle McKellar, "Louisiana's Hardest Working Historian," *New Orleans Times-Picayune,* March 14, 1926.

1. T. R. Johnson, "George Washington Cable", and Mary Ann Wilson, "Grace Klug,"

2. Bradbury, *Renaissance in the South,* 7–8.

3. Cammie's description of Shady Grove is in SB 18, 5, and 38.

4. Gutman and Frystak, "Carmelite 'Cammie' Garrett Henry," 2:175–95.

5. Ibid., 175.

6. Ibid., 185.

7. "Melrose Auction Sale, June 6 and 7, 1970"; Rubin, "Colfax Riot."

8. *Encyclopaedia Britannica Online,* s.v. "Historic Preservation: Safeguarding Treasures of the Past," by Jeannette L. Nolen, January 7, 2013, accessed November 5, 2016, https://www.britannica.com/topic/Historic-Preservation-Safeguarding-Treasures-of-the-Past-1904217.

9. Goodwin, *Bruton Parish Church Restored,* 33.

10. Megraw, *Confronting Modernity,* 34.

11. Dennison, "Summer Art Colonies in the South," 97–112.

12. Ellsworth Woodward, letter to Irma Sompayrac Willard, May 23, 1928, folder 8, Irma Sompayrac Willard Collection, CGHRC.

13. Lyle Saxon, "Art Colony Opens in Natchitoches," *New Orleans Times-Picayune,* June 12, 1921. This article is also pasted into SB 68, 116, CGHRC.

14. "Colony Closes at Natchitoches," *New Orleans Times-Picayune,* July 17, 1921.

15. "Natchitoches Art Colony Closes Summer Meeting," *New Orleans Times-Picayune,* July 9, 1922.

16. Mary Belle McKellar, "Log Cabin Studio Beautified," *Natchitoches Times,* April 16, 1926, SB 79, 78.

17. David Snell, "The Green World of Carrie Dormon," *Smithsonian,* February 1972. This article is in the Oliver Ford Collection, folder 8, CGHRC.

18. Gutman and Frystak, "Carmelite 'Cammie' Garrett Henry," 181.

19. Ada Jack Carver Snell, letter to Caroline Dormon, May 6, 1925, CDC, folder 519, CGHRC.

20. Lillian Trichel, letter to Cammie Henry, September 14, 1931, MC, folder 136, CGHRC.

21. Cammie Henry, letter to Caroline Dormon, August 16, 1920, CDC, folder 589. CGHRC.

22. "The Daily Program of Carrie Dormon" is found in SB 83, 23, CGHRC.

23. Cammie Henry, letter to Caroline Dormon, July 4, 1928, CDC, folder 597, CGHRC.

24. Caroline Dormon, letter to Cammie Henry, August 6, 1921, SB 83, 15, CGHRC.

25. Ibid.

26. Fran Holman Johnson, *Gift of the Wild Things,* 28.

27. Rawson, "Caroline Dormon," 121–39.

28. Caroline Dormon, letter to Cammie Henry, July 26, 1923, SB 83, 30. The Bessie Shaw Stafford article is found in the same scrapbook on p. 31.

29. In an August 15, 1923, letter to Cammie, Carrie said, "I may resign from the Dept. in Sept." Fran Holman Johnson confirms this was an accomplished fact in her biography, *Gift of the Wild Things,* 36.

30. Caroline Dormon, letter to Cammie Henry, May 6, 1922, SB 83, 26.

31. Chance Harvey, *Life and Selected Letters of Lyle Saxon,* 39–41.

32. Harvey describes Saxon's early journalism career ibid., 45–46.

33. King's quote is reproduced in Reed's *Dixie Bohemia* and originally appeared in the *New Orleans Times-Picayune* in March 1926.

34. FMCHC, folder 267, scan 65, August 31, 1940.

35. In *Plantation Memo,* François Mignon described Cammie and Lyle's first meeting at Grace King's salon (42). He said that "their first actual contact occurred at one of the 'Thursday Afternoons' at the home of Grace King."

36. Lyle Saxon, "Easter Sunday at Aunt Cammie's," April 22, 1923, *New Orleans Times-Picayune.* The article can be found pasted in SB 78, 40, CGHRC.

37. McKellar, "Louisiana's Hardest Working Historian."

38. In an undated, other than 1921, letter from Cammie Henry to Caroline Dormon, Cammie tells Carrie where to buy scrapbooks. CDC, folder 589.

39. This note is in SB 19, 125.

40. Cammie Henry, letter to François Mignon, n.d., FM, folder 38, CGHRC.

41. Megraw, *Confronting Modernity,* provides a thorough analysis of modernization through the arts in Louisiana, with a focus on the work of Lyle Saxon and Ellsworth Woodward, among others.

42. Cammie Henry, marginal note, SB 68, 106, CGHRC.

43. Diane M. Moore, "Mistress of Melrose," 11–12.

44. François Mignon recorded this comment in *Plantation Memo,* 42.

45. Lyle Saxon, letter to Cammie Henry, April 4, 1923, FWPC, folder 134, CGHRC.

46. Henry Chambers, letter to Cammie Henry, April 28, 1923, SB 81, 83, CGHRC.

47. Henry Chambers, letter to Cammie Henry, January 29, 1924, SB 81, 86, CGHRC.

48. McKellar, "Louisiana's Hardest Working Historian."

49. Henry Chambers, letter to Cammie Henry, November 20, 1925, SB 81, 93.

50. Henry Chambers, letter to Cammie Henry, January 10, 1928, SB 81, 95.

51. Mary Dell Fletcher provides a biographical sketch of Ada in which she discusses Ada's education and early association with Dormon, in *Collected Works of Ada Jack Carver,* 2.

52. Carver's story "The Stranger within the Gates" is referenced by Carver historian Donnis Ward Taylor in her dissertation "Louisiana's Literary Legacy," 12. Taylor notes that the story was published in the *Winnfield Guardian* in 1907. "The Ring" was published in the *New Orleans Item* in 1908, after Ada won a contest by the Louisiana Federation of Women's Clubs.

53. Fletcher, *Collected Works of Ada Jack Carver,* 2.

54. Ibid., 3. Fletcher notes that Snell was a "former faculty member" at the Normal School. For his thesis on Ada Jack Carver, Oliver Ford interviewed her sisters, Judith and Miriam, who confirmed that Ada met John Snell at Normal. His interview notes are in the Ford Collection, folder 5, CGHRC.

55. Fletcher, *Collected Works of Ada Jack Carver,* 3.

56. Ford, "Ada Jack Carver," 98–99.

57. Donnis Taylor says that Ada filled the tub with hot water, left the room, and the child fell into the water. She notes that another version of the story says a young servant drew the bath ("Louisiana's Literary Legacy," 16). There is no way to know for certain. The child's death certificate is in the Oliver Ford Collection, folder 1, CGHRC.

58. Taylor, "Louisiana's Literary Legacy," 16.

59. Ada Carver, note to Cammie Henry, MC, folder 1480, CGHRC.

60. Ada Jack Carver Snell, letter to Cammie Henry, August 17, 1924, MC, folder 78, CGHRC.

61. Ada Jack Carver, letter to Cammie Henry, August 31, 1924; both the poem and the note are found in MC, folder 323, CGHRC.

62. Ibid.

63. Ada Jack Carver Snell, letter to Cammie Henry, September 4, 1924, SB 217, 9.

64. Ibid. The story to which she refers in this letter is "Redbone," which was published in *Harper's,* February 1925.

65. Fletcher, *Collected Works of Ada Jack Carver,* 3.

66. Ada Jack Carver Snell, letter to Cammie Henry, May 1925, SB 83, 72.

67. Ada Jack Carver Snell, letter to Caroline Dormon, September 8, 1924, CDC, folder 518, CGHRC.

68. MacNeil, "It Doesn't Matter Where You Work."

3. THE MUSE OF MELROSE, 1925–1929

Epigraph: Caroline Dormon, "Gardening at Melrose," undated typescript, CDC, folder 762, CGHRC.

1. Cammie Henry, letter to Mary Debaillon, April 3, 1935, CDC, folder 608, CGHRC.

2. Lyle recorded events from this trip in his 1925 journal (MC, folder 133, CGHRC).

3. Thomas, *Lyle Saxon,* 43 and 50.

4. Reed, *Dixie Bohemia,* 101.

5. Chance Harvey, *Life and Selected Letters of Lyle Saxon,* 98.

6. Saxon, diary, 1925, MC, folder 133, CGHRC.

7. Ibid.

8. The receipt for the car is in SB 83, 41.

9. Saxon, diary, 1925, MC, folder 133, CGHRC.

10. Ibid. Also, a report of the accident appears in a Natchitoches newspaper clipping in an unnumbered scrapbook about Cammie Jr., 2, CGHRC.

11. Ford, "Ada Jack Carver," 123.

12. The newspaper account of the accident, dated March 21, 1925, is in the unnumbered Cammie Jr. scrapbook, 2.

13. Cammie Henry, letter to Caroline Dormon, 1925 (otherwise undated), CDC, folder 592, CGHRC.

14. Ada Jack Carver, letter to Caroline Dormon, undated other than 1925, CDC, folder 519, CGHRC.

15. Cammie Henry, letter to Caroline Dormon, April 6, 1938, CDC, folder 613, CGHRC.

16. Saxon's 1925 diary, MC, folder 133, CGHRC.

17. Ada Jack Carver, letter to Cammie Henry, April 5, 1925, SB 83, 67.

18. Cammie Henry, letter to Caroline Dormon, May 26, 1925, CDC, folder 591, CGHRC

19. Bradbury, *Renaissance in the South,* 10.

20. Mignon, *Plantation Memo,* 28.

21. MacDonald et al., "Archaeology of Local Myths and Heritage Tourism," 132.

22. Ibid., 126-42.

23. James Thomas describes Lyle's writing practice in his biography of Saxon. He says that Lyle would sit at a large desk he kept facing the blank walls "so that passersby would not disturb him; he wrote with his back to the windows" (*Lyle Saxon,* 128).

24. Ford makes reference to this writing exercise at Melrose in "Ada Jack Carver," 116.

25. Letter dated June 1925 to Cammie Henry, signed by Ada, Carrie, and David, SB 83, 73.

26. Ada Jack Carver, letter to Cammie Henry, June 13, 1925, SB 83, 80.

27. Ada Jack Carver, letter to Cammie Henry, October 1925, SB 83, 74.

28. Thomas, in *Lyle Saxon,* dates the story to 1925, citing a December 1925 letter from Lyle Saxon to Noel Straus.

29. Lyle Saxon, letter to Caroline Dormon, June 27, 1925, CDC, folder 555, CGHRC.

30. Lyle Saxon, letter to Caroline Dormon, August 7, 1925, CDC, folder 555, CGHRC.

31. Ford, "Ada Jack Carver," 116.

32. Quoted in Thomas, *Lyle Saxon,* 70.

33. Carrie's typed poem is found in CDC 975, CGHRC, publication notebook, along with many other of her unpublished poems.

34. Dormon, "Tree Studies in Blueprint." The article is found in SB 83, 32.

35. Dormon, "Blue Printing the Wildflowers." The article is found in SB 83, 35.

36. Caroline Dormon, letter to Lyle Saxon, n.d., CDC, folder 1164.

37. Caroline Dormon, letter to Cammie Henry, June 2, 1926, SB 83, 124.

38. Cammie Henry, letter to Caroline Dormon, January 14, 1925, CDC, folder 591, CGHRC.

39. Cammie Henry, letter to Caroline Dormon, May 26, 1925, CDC, folder 591, CGHRC.

40. Lyle Saxon, letter to Caroline Dormon, June 27, 1925, CDC, folder 555, CGHRC.

41. Cammie Henry, letter to Caroline Dormon, January 26, 1928, CDC, folder 597, CGHRC.

42. Alberta Kinsey's comments on Cammie Henry are from a typed statement composed after Cammie's death and are found in folder 41, FM, CGHRC.

43. Cammie Henry, letter to Caroline Dormon, July 6, 1925, CDC, folder 592, CGHRC.

44. Cammie Henry, letter to Caroline Dormon, July 16, 1925, CDC, folder 592, CGHRC.

45. Lyle Saxon, letter to Caroline Dormon, August 7, 1925, CDC, folder 555, CGHRC.

46. Ibid.

47. The washday photos are in SB 79, 46-57.

48. Lyle Saxon, "Ada Jack Carver's Prize-Winning Play to Be Placed in Belasco Cup Contest," *New Orleans Times-Picayune,* February 21, 1926, SB 83, 76, CGHRC.

49. Ibid.

50. Lyle Saxon, letter to Caroline Dormon, August 7, 1925, CDC, folder 555, CGHRC.

51. Cammie Henry, letter to Caroline Dormon, August 25, 1925, CDC, folder 592, CGHRC.

52. Cammie Henry, letter to Caroline Dormon, September 10, 1925, CDC, folder 592, CGHRC.

53. Cammie Henry, letter to Caroline Dormon, October 29, 1925, CDC, folder 592, CGHRC.

54. Ada Jack Carver, letter to Cammie Henry, September 26, 1925, SB 83, 73.

55. Ada Jack Carver, letter to Cammie Henry, October 11, 1925, SB 83, 74.

56. Cammie's marginal notes are on a letter from Ada Jack Carver to Cammie dated November 4, 1925, SB 83, 74, CGHRC.

57. Lyle Saxon, letter to Cammie Henry, n.d., FWPC, folder 125, CGHRC.

58. Cammie Henry, letter to Caroline Dormon, January 21, 1926, CDC, folder 593, CGHRC.

59. Gutman and Frystak, "Carmelite 'Cammie' Garrett Henry," 188; see also Scarborough, *On the Trail of Negro Folk-Songs*, 18, for details on her visit with Uncle Israel and Aunt Jane.

60. Chance Harvey, *Life and Selected Letters of Lyle Saxon*, 139.

61. Thomas, *Lyle Saxon*, 126.

62. McKellar, "Louisiana's Hardest Working Historian," SB 79, 63.

63. Cammie Henry, letter to Caroline Dormon, August 11, 1922, CDC, folder 589, CGHRC.

64. Cammie Henry, letter to Caroline Dormon, March 25, 1926, CDC, folder 593, CGHRC.

65. Caroline Dormon, letter to Cammie Henry, 1925 (otherwise undated), SB 83, 125.

66. Fran Holman Johnson, *Gift of the Wild Things*, 47.

67. Cammie described the quilt she received from Madame Roque in SB 217, 58.

68. Mrs. Bowman Whited, letter to Cammie Henry, May 26, 1926, SB 217, 46.

69. The results of the contest are found in SB 217, 28; one article is dated May 9, 1926, and another May 10, 1926. Both are from unidentified newspapers.

70. Irma Sompayrac, telegram to Cammie Henry, May 5, 1926, SB 83, 77.

71. Cammie Henry, letter to Caroline Dormon, June 12, 1926, CDC, folder 593, CGHRC.

72. Caroline Dormon, letter to Cammie Henry, April 28, 1926, SB 83, 46.

73. Ada Jack Carver, letter to Cammie Henry, July 25, 1926, SB 217, 53.

74. Caroline Dormon, letter to Cammie Henry, August 14, 1926, SB 83, 125.

75. Ibid.

76. Caroline Dormon, letter to Cammie Henry, August 26, 1926, SB 83, 127.

77. Ada Jack Carver, letter to Cammie Henry, n.d. [other than "Friday"], MC, folder 324, CGHRC.

78. Ibid.

79. Caroline Dormon, letter to Cammie Henry, November 2, 1926, MC, folder 326, CGHRC.

80. Babb, *My Sketchbook*, 50–52.

81. Ibid., 81.

82. Ibid., 84.

83. Ibid., 87.

84. Arthur Babb, letter to Cammie Henry, December 11, 1944, SB 79, 127.

85. Cammie Henry, letter to Caroline Dormon, December 6, 1926, CDC, folder 594, CGHRC.

86. Lyle Saxon, letter to Cammie Henry, December 20, 1926, SB 220, 132.

87. Lillian Trichel's typescript of "Wild Goose Chase" is found in MC, folder 327, CGHRC.

88. Chance Harvey, *Life and Selected Letters of Lyle Saxon*, 127.

89. Cammie Henry, letter to Caroline Dormon, April 25, 1927, MC, folder 595, CGHRC.

90. There are pages of photos Lyle took while covering the flood in SB 220, specifically 123–29.

91. Lyle's 1927 diary is in MC, folder 133, CGHRC.

92. Chance Harvey, *Life and Selected Letters of Lyle Saxon*, 248.

93. Ibid., 247.

94. Lyle Saxon, note to Cammie Henry, August 8, 1927, SB 225, 25.

95. Chance Harvey, *Life and Selected Letters of Lyle Saxon*, 249.

96. It should be noted that Cammie certainly knew proper grammar, but on occasion she and Carrie would speak this way with each other in fun. Cammie Henry, letter to Caroline Dormon, November 2, 1927, CDC, folder 596, CGHRC.

97. Cleveland B. Chase's *New York Times* review, dated November 27, 1927, is in SB 220, 76.

98. Titzell's remarks are attached to a copy of the review that he sent to Cammie, SB 220, 98.

99. Bernard DeVoto's *Saturday Evening Post* review is in SB 220, 94.

100. Thomas, *Lyle Saxon*, 96.

101. Parrish, *Flood Year,* 182.

102. "Lyle Saxon Back from New York to Write Novel," *Baton Rouge State-Times,* January 3, 1930, in SB 12, 114.

103. Robert Thomas Hardy, letter to Caroline Dormon, June 25, 1925, CDC, folder 920, CGHRC.

104. Caroline Dormon, letter to Ann Titzell, July 1, 1947, CDC, folder 486, CGHRC.

105. Cammie Henry, letter to Caroline Dormon, August 10, 1927, CDC, folder 596, CGHRC.

106. Caroline Dormon, letter to Cammie Henry, September 11, 1927, SB 83, 46.

107. Lyle Saxon, letter to Caroline Dormon, August 29, 1927, CDC, folder 556, CGHRC.

108. Cammie Henry, note, SB 83, 47.

109. Caroline Dormon, "Gardening at Melrose," typescript, n.d., CDC, folder 762, CGHRC.

110. Lyle Saxon, letter to Caroline Dormon, June 1, 1928, CDC, folder 556, CGHRC.

111. Mary Daggett Lake, "A Pilgrimage to the Heart of Old Louisiana," *Fort Worth Star-Telegram,* September 2, 1926, in SB 79, 106.

112. Cammie Henry, note, SB 68, 47.

113. Mignon, *Plantation Memo,* 15.

114. Cammie Henry, note, SB 68, 47.

115. Cammie Henry, note, SB 83, 68.

116. Ada Whitfield Jack Carver, letter to Cammie Henry, November 17, 1929, MC, folder 326, CGHRC.

117. Caroline Dormon, letter to Cammie Henry, December 1929, SB 83, 133.

118. Gilbert Mead's review of *Old Louisiana* is found in SB 12, 113.

4. A WIDENING INFLUENCE, 1930–1936

Epigraph: Harold M. Case, "Melrose—Mecca of the Muse—Jewel of Cane River," *Tensas Gazette,* June 29, 1933, MC, folder 452, CGHRC.

1. Saxon's 1930 diary, MC, folder 1223, CGHRC.

2. Mignon, *Plantation Memo*, 45.

3. Chance Harvey, *Life and Selected Letters of Lyle Saxon*, 110.

4. Ibid., 156.

5. Joy Jackson, "Prohibition in New Orleans," 261–84.

6. Blotner, *Faulkner*, 127.

7. Henry Jack, letter to Lyle Saxon, February 14, 1927, MC, folder 82, CGHRC.

8. Lyle Saxon, note to Cammie Henry, n.d., FWPC, folder 125, CGHRC.

9. Lyle Saxon, letter to Caroline Dormon, June 15, 1934, CDC, folder 560, CGHRC.

10. Ada Jack Carver, letter to Cammie Henry, October 28, 1929, SB 217, 68.

11. Ada Jack Carver, letter to Cammie Henry, n.d., SB 83, 71.

12. Lyle Saxon's 1930 diary, MC, folder 1223, CGHRC.

13. Cammie Henry, note, SB 217, 79.

14. Ada Jack Carver, letter to Cammie Henry, February 19, 1930, SB 217, 79.

15. Leudivine Erwin Garrett, letter to Caroline Dormon, March 9, 1930, CDC, folder 600, CGHRC.

16. Thomas, *Lyle Saxon*, 126.

17. Saxon, diary entry, March 13, 1930, MC, folder 1223, CGHRC.

18. Saxon, diary entry, March 14, 1930, MC, folder 1223, CGHRC.

19. Lyle Saxon, letter to Cammie Henry Jr., April 6, 1931, SB 203, 67.

20. Saxon, diary entry, March 17, 1930, MC, folder 1223, CGHRC.

21. Saxon, diary entry, March 19, 1930, MC, folder 1223, CGHRC.

22. Saxon, diary entry, March 16, 1930, MC, folder 1223, CGHRC.

23. Cammie Henry, letter to Caroline Dormon, December 15, 1930, CDC, folder 601, CGHRC.

24. Caroline Dormon, letter to Cammie Henry, December 18, 1930, MC, folder 337, CGHRC.

25. Saxon, diary entry, April 16, 1930, MC, folder 1223, CGHRC.

26. Lyle Saxon, letter to Caroline Dormon, September 15, 1930, CDC, folder 557, CGHRC.

27. Dormon, "Southern Personalities." The article can be found in SB 213, 5.

28. Caroline Dormon, letter to Cammie Henry, December 30, 1930, MC, folder 337, CGHRC.

29. Lyle Saxon, letter to Caroline Dormon, July 21, 1930, CDC, folder 557, CGHRC.

30. Lyle Saxon, letter to Caroline Dormon, September 14, 1930, CDC, folder 557, CGHRC.

31. Ibid.

32. Faust, "History of Malaria."

33. Cammie Henry, letter to Caroline Dormon, August 7, 1930, CDC, folder 601, CGHRC.

34. Cammie Henry, letter to Caroline Dormon, August 16, 1930, CDC, folder 601, CGHRC.

35. Lyle Saxon, letter to Caroline Dormon, September 14, 1930, CDC, folder 557, CGHRC. Saxon's birthday was September 4.

36. Cammie Henry Jr., letter to her mother, October 17, 1930. In a series of letters home written through that period, Cammie Jr. asks about her brother Dan, sister-in-law Celeste, Melrose help Massoline, and Mary, and Fred Wilson. SB 203, 90.

37. Caroline Dormon, letter to Cammie Henry Jr., November 1930, SB 203, 80.

38. A statement from Salem Academy and College dated September 10, 1930, is in SB 203, 28. In a letter to Carrie from September 2, 1931, Cammie wrote, "I borrowed of my inheritance from Mother to send Sister back to Salem," although it is not clear what Leudivine would have had to leave given that she was in poor financial shape after her husband's death. CDC, folder 603, CGHRC.

39. Cammie Henry, letter to Caroline Dormon, September 27, 1930, CDC, folder 601, CGHRC.

40. Ibid.

41. Cammie Henry, letter to Caroline Dormon, November 28, 1930, CDC, folder 601, CGHRC.

42. Cammie Henry, letter to Caroline Dormon, December 4, 1930, CDC, folder 601, CGHRC.

43. Robina Denholm, letter to Caroline Dormon, November 11, 1935, CDC, folder 527, CGHRC.

44. Robina Denholm, letter to Caroline Dormon, undated but marked 1941–43 by archivist, CDC, folder 530, CGHRC.

45. Leudivine Erwin Garrett, letter to Caroline Dormon, January 4, 1931, CDC, folder 602, CGHRC.

46. Harold Case, "Melrose—Mecca of the Muse—Jewel of Cane River—Still Keeps Open House for Writers and Artists of Nation," *Tensas Gazette,* 1933. This article is found in SB 83, 7, and is the second article in a series Case wrote about Melrose.

47. Otto Claitor, letter to Cammie Henry, January 30, 1931, MC, folder 41, CGHRC.

48. Otto Claitor, letter to Cammie Henry, June 29, 1931, MC, folder 41, CGHRC.

49. Ada Jack Carver, letter to Cammie Henry, March 27, 1931, SB 217, 81.

50. Cammie Henry, note, SB 83, 79.

51. Cammie Henry, letter to Caroline Dormon, June 11, 1931, CDC, folder 601, CGHRC.

52. Irene Wagner, interview by Kathryn Bridges, Bailey Collection, Northwestern State University of Louisiana, Natchitoches.

53. Alfonso Lahcar, "In the Land of Romance," *Shreveport Times,* March 6, 1931. This article is found in SB 79, 130.

54. Cammie Henry, diary entry, April 13, 1934, FM, bound vol. 182, CGHRC.

55. Cammie Henry, diary entry, May 30, 1934, FM, bound vol. 182, CGHRC.

56. Cammie Henry, letter to Caroline Dormon, March 10, 1931, CDC, folder 602, CGHRC.

57. Lyle Saxon, letter to Caroline Dormon, 1931 (otherwise undated), CDC, folder 557, CGHRC.

58. Thomas, *Lyle Saxon,* 126.

59. Ibid.

60. Chance Harvey, *Life and Selected Letters of Lyle Saxon,* 170.

61. Thomas, "Lyle Saxon's Struggle."

62. Lyle Saxon, letter to Cammie Henry Jr., April 6, 1931, SB 203, 67.

63. Ibid.

64. The article and the note can be found in SB 12, 74.

65. Cammie collected several articles surrounding the controversy of *Cane Juice,* which are in SB 12, 36–46, and include articles from the *Shreveport Times,* the *Shreveport Journal,* the *New Orleans Item,* and the *Baton Rouge State-Times.*

66. John Earle Uhler, letter to Cammie Henry, October 13, 1931, SB 12, 42.

67. Lyle Saxon, letter to Caroline Dormon, May 28, 1931, CDC, folder 557, CGHRC.

68. Doris Ulmann, letter to Cammie Henry, May 31, 1931, MC, folder 137, CGHRC.

69. Cammie Henry, letter to Caroline Dormon, May 28, 1931, CDC, folder 633, CGHRC.

70. Doris Ulmann, letter to Lyle Saxon, June 14, 1934, FWPC, folder 121, CGHRC.

71. Jacobs, *Life and Photography of Doris Ulmann*, is the definitive biography of Ulmann.

72. Cammie Henry, letter to Caroline Dormon, August 10, 1931, CDC, folder 603, CGHRC.

73. Cammie Henry, letter to Caroline Dormon, August 26, 1931, CDC, folder 603, CGHRC.

74. Ibid.

75. Leudivine Erwin Garrett, letter to Caroline Dormon, October 28, 1931, CDC, folder 603, CGHRC.

76. Cora Bristol Nelson, letter to Cammie Henry, March 2, 1931, MC, folder 1289, CGHRC.

77. Cora Bristol Nelson, letter to Cammie Henry, n.d., MC, folder 1289, CGHRC.

78. Ibid.

79. Cammie Henry, marginal notation on a July 13, 1931, letter from Cora Bristol Nelson, MC, folder 1289, CGHRC.

80. *John H. Henry v. Robert N. Henry,* May 26, 1941, 10th Judicial District Court, Natchitoches, La., no. 24055.

81. Quoted in "Death of Grace King," *Louisiana Historical Quarterly* 15 (1932): 334.

82. Cammie Henry, letter to Caroline Dormon, n.d., CDC, folder 624, CGHRC.

83. Chance Harvey, *Life and Selected Letters of Lyle Saxon,* 165.

84. Harvey discusses Saxon's health ibid., 165; Thomas quotation in *Lyle Saxon,* 51.

85. Lyle Saxon, letter to Caroline Dormon, July 22, 1932, CDC, folder 557, CGHRC.

86. Cammie Henry, letter to Caroline Dormon, September 18, 1932, CDC, folder 605, CGHRC.

87. Lyle Saxon, letter to Caroline Dormon, October 13, 1932, CDC, folder 557, CGHRC.

88. Cammie Henry, letter to Caroline Dormon, November 10, 1932, CDC, folder 605, CGHRC.

89. Sarah Jackson's biography of Karle Wilson Baker, *Texas Woman of Letters,* was published in 2005.

90. Cammie Henry, letter to Caroline Dormon, March 22, 1933, CDC, folder 606, CGHRC.

91. Cammie Henry, letter to Caroline Dormon, March 25, 1933, CDC, folder 606, CGHRC.

92. "Minden Merchant Gets Bird's Eye View of Tornado," *Shreveport Times,* dated by Cammie as May 1933, SB 217, 90.

93. Mary Belle McKellar, letter to Cammie Henry, 1933 (otherwise undated), MC, folder 97, CGHRC.

94. Lyle Saxon, letter to Caroline Dormon, October 8, 1933, CDC, folder 558, CGHRC.

95. Lyle Saxon, letter to Caroline Dormon, August 21, 1933, CDC, folder 558, CGHRC.

96. Ibid.

97. Cammie Henry, letter to Caroline Dormon, August 23, 1933, CDC, folder 607, CGHRC.

98. Lyle Saxon, letter to Caroline Dormon, November 30, 1933, CDC, folder 558, CGHRC.

99. Cammie Henry, diary entry, December 11, 1934, FM, bound vol. 182, CGHRC.

100. Cammie Henry, letter to Cammie Henry Jr., March 19, 1934, MC, folder 1453, CGHRC.

101. Cammie Henry, diary entry, May 26, 1934, FM, bound vol. 182, CGHRC.

102. Cammie Henry, diary entry, December 27, 1934, FM, bound vol. 182, CGHRC.

103. Cammie Henry, diary entry, October 2, 1934, FM, bound vol. 182, CGHRC.

104. Cammie Henry, diary entry, October 22, 1934, FM, bound vol. 182, CGHRC.

105. Cammie Henry, diary entry, October 30, 1934, FM, bound vol. 182, CGHRC.

106. Cammie Henry, diary entry, October 21, 1934, FM, bound vol. 182, CGHRC.

107. Delphin, *Boots in the Grass,* 95, describes Bubba's location on Highway 119 at Melrose and its reputation as a "community juke joint."

108. Cammie Henry, diary entry, November 1, 1934, FM, bound vol. 182, CGHRC.

109. Rudolph Fuchs, "Mrs. Cammie Henry and Her Interest in Handweaving," typescript, n.d., MC, folder 41, CGHRC.

110. Cammie Henry, diary entry, November 27, 1934, FM, bound vol. 182, CGHRC.

111. Cammie Henry, diary entries, December 6 and 7, 1934, FM, bound vol. 182, CGHRC.

112. Lewis, "Mr. Feuille," 360–61.

113. The receipt for the art restoration is in MC, folder 10; correspondence between Saxon and Cammie Henry about the painting is in FWPC, folder 134, CGHRC.

114. Cammie Henry, diary entry, May 7, 1934, FM, bound vol. 182, CGHRC.

115. Charles Mazurette, letter to Cammie Henry, December 4, 1933, MC, folder 99, CGHRC.

116. Charles Mazurette, letter to Cammie Henry, December 9, 1933, MC, folder 99, CGHRC.

117. Charles Mazurette, letter to Cammie Henry, January 6, 1934, MC, folder 99, CGHRC.

118. Charles Mazurette, letter to Cammie Henry, June 27, 1934, MC, folder 99, CGHRC.

119. Cammie Henry, diary entry, May 12, 1934, FM, bound vol. 182, CGHRC.

120. Cammie Henry, letter to Caroline Dormon, March 25 and April 26, 1926, CDC, folder 593, CGHRC.

121. Lyle Saxon, letter to Caroline Dormon, November 30, 1933, CDC, folder 558, CGHRC.

122. Cammie Henry, diary entry, May 22, 1934, FM, bound vol. 182, CGHRC.

123. Lyle Saxon, letter to Caroline Dormon, May 22, 1934, CDC, folder 559, CGHRC.

124. Cammie Henry, diary entry, October 5, 1934, FM, bound vol. 182, CGHRC.

125. Cammie Henry, diary entry, November 13, 1934, FM, bound vol. 182, CGHRC.

126. Cammie Henry, diary entry, December 23, 1934, FM, bound vol. 182, CGHRC.

127. Cammie Henry, letter to Caroline Dormon, February 3, 1935, CDC, folder 608, CGHRC.

128. Lyle Saxon, letter to Caroline Dormon, February 6, 1935, CDC, folder 561, CGHRC.

129. Edith Wyatt Moore, "In Memory of Louisiana's Grand Old Lady," *Tensas Gazette,* clipping in SB 80, 73.

130. Lillian Hall Trichel, typed comments about Cammie Henry, July 11, 1950, FM, folder 72, CGHRC.

131. Cammie Henry, letter to Caroline Dormon, February 12, 1935, CDC, folder 608, CGHRC.

132. Letter from "The Editors" at *Harper's* to Cammie Henry, August 2, 1935, SB 217, 97.

133. Lillian Trichel, letter to Cammie Henry, August 11, 1936, MC, folder 136A, CGHRC.

134. A large part of the correspondence between Cammie Henry and Debora Abramson is in MC, folder 95, which covers five years.

135. The letter is found in MC, folder 169, CGHRC.

5 A SHIFT IN COURSE, 1937–1940

Epigraph: Cammie Henry, letter to François Mignon, January 10, 1939, FM, folder 23, CGHRC.

1. Chance Harvey, *Life and Selected Letters of Lyle Saxon*, 180.

2. Jo Thompson, "Lyle Saxon, Seasoned Reporter, Likes Reporting Far More than Writing Books—but Writes Books," *Baton Rouge State-Times*, August 5, 1937, 3. This article is in SB 225, 86.

3. Thomas, *Lyle Saxon*, 125.

4. George Stevens, "A Poignant Story of the 'Free Mulattoes,'" *Saturday Review of Literature*, July 10, 1937, 19. This review is in SB 223, 25.

5. Edward Larocque Tinker, "A Dramatic Novel of Louisiana," *New York Times Book Review*, July 11, 1937, 1. This review is found in SB 223, 26.

6. Reed, *Dixie Bohemia*, 116–18.

7. Irby C. Nichols, "Announcement of Natchitoches Meetings March 13, 14," *Mathematics News Letter* 5, no. 5 (January 1931): 6, JSTOR (3027931).

8. Caroline Dormon, diary entry, July 20, 1937, CDC 977, CGHRC.

9. Fran Holman Johnson, *Gift of the Wild Things*, 2

10. Cora Bristol Nelson, letter to Cammie Henry, September 20, 1937, MC, folder 1290, CGHRC.

11. Julia Peterkin, letter to Cammie Henry, undated, SB 225, 106.

12. Caroline Dormon, letter to Cammie Henry, February 20, 1938, SB 83, 19.

13. Fran Holman Johnson, *Gift of the Wild Things*, 79–87.

14. Ford, "François Mignon," 57.

15. Cammie Henry, letter to François Mignon, September 30, 1938, FM, folder 22, CGHRC.

16. Cammie Henry, letter to François Mignon, July 31, 1939, FM, folder 32, CGHRC.

17. Cammie Henry, letter to François Mignon, January 10, 1939, FM, folder 23, CGHRC.

18. Cammie Henry, letter to François Mignon, November 12, 1938, FM, folder 22, CGHRC.

19. Cammie Henry, letter to François Mignon, January 10, 1939, FM, folder 23, CGHRC.

20. Cammie Henry, letter to François Mignon, January 21, 1939, FM, folder 23, CGHRC.

21. Cammie Henry, letter to François Mignon, December 31, 1938, FM, folder 23, CGHRC.

22. Cammie Henry, letter to François Mignon, December 31, 1939, FM, folder 23, CGHRC.

23. Cammie Henry, letter to François Mignon, February 5, 1939, FM, folder 24, CGHRC.

24. Lyle Saxon, letter to Caroline Dormon, January 26, 1939, CDC, folder 564, CGHRC.

25. Ford, "Ada Jack Carver," 141.

26. Cammie Henry, letter to the principal of Shady Grove High School, January 15, 1939, SB 80, 94.

27. Eddie Dryer, letter to Robina Denholm, dated March 13, 1939, CDC, folder 564, CGHRC.

28. Robina Denholm, letter to François Mignon, August 20, 1939, FM, folder 34. CGHRC.

29. Ford, "François Mignon," 51–59.

30. François Mignon, journal entry, October 28, 1939, folder 260, scan 3. FMCHC.

31. Mignon's description of the plants is found in his 1939 journal, folder 266, scan 41; the description of the gin operations is in folder 268, scan 37, FMCHC.

32. François Mignon, journal entry, November 8, 1939, folder 260, scan 18, FMCHC.

33. Ibid., scan 20.

34. François Mignon, journal entry, November 13, 1939, folder 260, scan 42, FMCHC.

35. François Mignon, journal entry, November 17, 1939, folder 260, scan 54, FMCHC.

36. François Mignon, journal entry, November 24, 1939, folder 260, scan 72, FMCHC.

37. There has been some confusion at times between Cammie Henry and her daughter; in Mary Lyons's 1998 children's book about Clementine Hunter, *Talking with Tebé,* Lyons indicates that Cammie Jr. was Hunter's employer, but that was never the case. Lyons quotes Cammie Jr.'s son, saying, "Her son recalls she was an alcoholic whose moods ran hot and cold." While Lyons does not cite the source with any further detail, this would have been most likely in reference to Cammie Jr. rather than her mother, who seldom drank and avoided alcohol, due in part perhaps to some reported alcoholism on both the Erwin and Garrett sides of the family.

38. François Mignon, journal entry, November 14, 1939, folder 260, scans 47–48, FMCHC.

39. François Mignon, journal entry, November 20, 1939, folder 260, scan 60, FMCHC.

40. François Mignon, journal entry, November 22, 1939, folder 260, scan 64, FMCHC.

41. François Mignon, journal entry, January 11, 1940, folder 261, scan 23, FMCHC.

42. François Mignon, journal entry, November 22, 1939, folder 260, scan 65.

43. François Mignon, journal entry, December 19, 1939, folder 118, scan 260, FMCHC.

44. Ibid. The story of Clementine Hunter and her painting has been told in several excellent works, notably Shiver and Whitehead, *Clementine Hunter.*

45. Francois Mignon, journal entry, December 25, 1939, folder 260, scan 128, FMCHC.

6. Cultivating the Legacy, 1940–1948

Epigraph: François Mignon, journal entry, November 16, 1939, folder 260, scan 58, FMCHC.

1. François Mignon, journal entry, May 7, 1940, folder 264, scan 13, FMCHC.

2. Jeff Matthews, "Remembering the Louisiana Maneuvers," *Town Talk,* July 29, 2016, accessed October 27, 2016, http://www.thetowntalk.com/story/news/2016/07/29/remembering-louisiana -maneuvers/87575988/.

3. François Mignon, journal entry, May 17, 1940, folder 264, scan 42, FMCHC.

4. François Mignon, journal entry, April 17, 1940, folder 263, scan 23, FMCHC.

5. François Mignon, journal entry, March 2, 1940, folder 262, scan 45, FMCHC.

6. Henry Tyler, letter to Cammie Henry, July 22, 1936, MC, folder 457, CGHRC.

7. François Mignon, journal entry, January 10, 1940, folder 261, scan 17, FMCHC.

8. François Mignon, journal entry, April 18, 1940, folder 263, scan 24, FMCHC.

9. Patrick's thesis is titled "Literature in the Louisiana Plantation Home prior to 1861: A Study in Literary Culture," 1937, a copy of which is in the CGHRC.

10. François Mignon, journal entry, July 1, 1940, folder 260, scan 1, FMCHC.

11. François Mignon, journal entry, July 6, 1940, folder 260, scan 10, FMCHC.

12. Lillian Trichel, letter to Caroline Dormon, May 27, 1940, CDC, folder 694. CGHRC.

13. Cammie Henry, letter to Caroline Dormon, January 23, 1941, CDC, folder 616. CGHRC.

14. Mary Belle McKellar, letter to Cammie Henry, January 9, 1934. MC, folder 327. CCHRC.

15. Ford, "Ada Jack Carver," 196

16. This note is found in an unnumbered scrapbook for Cammie Jr. at NSU, CGHRC.

17. Letter from "Clif," December 6, 1941, SB 217, 83.

18. Taylor, "Louisiana's Literary Legacy," 43.

19. Ada Jack Carver, letter to Caroline Dormon, July 12, 1925, CDC, folder 529, CGHRC.

20. Ada Jack Carver, letter to Irma Sompayrac Willard, n.d., quoted in Taylor, *Louisiana Literary Legacy*, 41.

21. Joe Henry, letter to Cammie Henry, January 28, 1942, SB 79, 114.

22. François Mignon, journal entry, July 19, 1941, folder 266, scan 41, FMCHC.

23. Lyle Saxon, letter to Caroline Dormon, 1942 (otherwise undated), CDC, folder 1164, CGHRC.

24. Lyle Saxon's mock letter is dated July 17, 1943, and is in MC, folder 1241, CGHRC.

25. This description of Melrose is from a letter Saxon wrote to New Orleans attorney Judith Hyams Douglas, in Chance Harvey, *Life and Selected Letters of Lyle Saxon*, 207.

26. Cammie Henry, letter to Caroline Dormon, March 29, 1943, CDC, folder 618, CGHRC.

27. Cammie Henry, letter to Caroline Dormon, May 6, 1943, CDC, folder 618, CGHRC.

28. Harnett Kane, letter to Cammie Henry, July 16, 1945, MC, folder 456, CGHRC.

29. François Mignon, letter to Caroline Dormon, October 31, 1945 CDC, folder 1462, CGHRC.

30. Cammie Henry, letter to Irene Wagner, April 6, 1946, Irene Wagner Collection, folder 8, CGHRC.

31. Kane, *Plantation Parade*, 271–72.

32. Ibid., 281.

33. Irene Wagner, interviewed by Kathryn Bridges, March 5, 1978, Bailey Collection, Northwestern State University of Louisiana, Natchitoches.

34. Frances Benjamin Johnston, letter to Cammie Henry, March 28, 1946, MC folder 84, CGHRC.

35. François Mignon journal entry, December 10, 1948, FMCHC.

36. Ibid.

37. Cammie Henry, note, SB 220, 136.

38. Lyle Saxon, letter to Cammie Henry, October 16, 1945, MC, folder 339, CGHRC.

39. Thomas, *Lyle Saxon*, 175.

40. François Mignon, journal entry, April 9, 1946, FMCHC.

41. Robert Tallant, letter to Cammie Henry, April 14, 1946, FWPC, folder 148, CGHRC.

42. Alberta Kinsey, letter to Cammie Henry, April 13, 1946, MC, folder 333, CGHRC.

43. Charles Mazurette, letter to Cammie Henry, April 12, 1946, MC, folder 333, CGHRC.

44. Cammie Henry, letter to Essie Mae Culver, Louisiana Library Commission, April 20, 1946, Joseph M. Henry Collection, folder 4, CGHRC.

45. Cammie Henry, note, SB 225, 1.

46. François Mignon, journal entry, April 14, 1948, FMCHC.

47. Mignon, *Plantation Memo*, 315.

48. A copy of Cammie Henry's death certificate is in the Joseph M. Henry Collection, folder 5, CGHRC.

49. François Mignon, journal entry, November 20, 1948, FMCHC.

50. François Mignon, journal entry, November 18, 1948, FMCHC

51. Ibid.

52. Ibid.

EPILOGUE

Epigraph: Lyle Saxon, letter to Cammie Henry, n.d., MC, folder 132, CGHRC.

1. Winston De Ville, appraisal dated June 22, 1970, SGHC, box 5, folder 41, CGHRC.

2. Fletcher, *Collected Works of Ada Jack Carver,* 188.

3. Ford, "Ada Jack Carver," 142.

4. François Mignon, letter to Caroline Dormon, February 25, 1949, CDC, folder 1463.

5. Visitation numbers from the Association for the Preservation of Historic Natchitoches, "Melrose Plantation Long-Range Interpretive Plan," 2013.

6. Lyle Saxon, "Easter Sunday at Aunt Cammie's," *New Orleans Times-Picayune,* April 22, 1923. The article is pasted in SB 78, 40, CGHRC.

BIBLIOGRAPHY

UNPUBLISHED LETTERS, SCRAPBOOKS, AND MEMORABILIA

Cammie Garrett Henry Research Center, Northwestern State University of Louisiana
    Association for the Preservation of Historic Natchitoches Collection.
    Carver, Hampton. Collection.
    Dormon, Caroline. Collection.
    Federal Writers Project Collection.
    Ford, Oliver. Collection.
    Henry, Joseph M. Collection.
    Henry, Stephen G. Collection.
    Melrose Collection.
    Mignon, François. Collection.
    Scrapbooks compiled by Cammie Garrett Henry.
    Wagner, Irene. Collection.
    Wenk, Eugene Lloyd. Collection.
    Willard, Irma Sompayrac. Collection.
Mignon, François. Journal. 1948. Southern Historical Collection, Louis Round Wilson
    Special Collections Library, University of North Carolina, Chapel Hill.
Taylor, Miles, and Family Papers. Hill Memorial Library, Louisiana State University Libraries, Baton Rouge.

INTERVIEWS

Sylvie, Alexander. Telephone interview with author, November 5, 2016.
Wagner, Irene. Interview by Kathryn Bridges on Cammie G. Henry, n.d. MP3 audio.
    Bailey Collection Tapes regarding Clementine Hunter and Melrose, Northwestern

State University of Louisiana Library. https://library.nsula.edu/clementine-hunter
    -and-melrose-tapes/.
Wenk, John. Telephone interview with author, May 11, 2015.

                        BOOKS, ARTICLES, AND REPORTS

Allured, Janet. *Louisiana Women: Their Lives and Times.* Athens: University of Georgia
    Press, 2009.
Andreassen, John C. L. "Frances Benjamin Johnston and Her Views of Uncle Sam." *Louisi-
    ana History* 1, no. 2 (Spring 1960): 130–36. Accessed January 14, 2017. http://www.jstor
    .org/stable/4230558.
Appleton, Thomas H., and Angela Boswell. *Searching for Their Places: Women in the South
    across Four Centuries.* Columbia: University of Missouri Press, 2003.
Babb, Arthur. *My Sketchbook: Memoirs of Natchitoches and Plaquemine.* Edited by Neill
    Cameron. Natchitoches, La.: Northwestern State University Press, 1996.
Barry, John M. *The Great Influenza: The Story of the Deadliest Pandemic in History.* New
    York: Penguin Books, 2005.
Biographical and Historical Memoirs of Northwest Louisiana. Nashville: Southern Pub. Co., 1890.
Blotner, Joseph L. *Faulkner: A Biography.* Jackson: University of Mississippi Press, 1974.
Bradbury, John M. *Renaissance in the South: A Critical History of the Literature, 1920–1960.*
    Chapel Hill: University of North Carolina Press, 1963.
Campanella, Richard. *Lost New Orleans.* London: Pavilion Books, 2015.
Carver, Ada Jack. *The Collected Works of Ada Jack Carver.* Edited by Mary Dell Fletcher.
    Natchitoches, La.: Northwestern State University Press, 1980.
Clement, William E. *Plantation Life on the Mississippi.* New Orleans: Pelican Publishing
    Co., 1952.
Cox, Karen L. *Dixie's Daughters: The United Daughters of the Confederacy and the Preserva-
    tion of Confederate Culture.* Gainesville: University Press of Florida, 2003.
Crespi, Muriel (Miki). "A Brief Ethnography of Magnolia Plantation: Planning for Cane
    River Creole National Historic Park. Archeology and Ethnography Program." Na-
    tional Center for Cultural Resources. Washington, D.C.: National Park Service,
    2004. Accessed March 31, 2016. https://www.nps.gov/archeology/pubs/studies/
    study04A.htm.
Davis, Jack. *Race against Time: Culture and Separation in Natchez since 1930.* Baton Rouge:
    Louisiana State University Press, 2001.
Delphin, Walter M. *Boots in the Grass: Story of Love and War.* Natchitoches, La.: Natchi-
    toches Art Blend, 2015.

Dennison, Mariea Caudill. "Summer Art Colonies in the South, 1920–1940." In *Cross-Roads. A Southern Culture Annual 2004.* Edited by Ted Olson, 97–112. Macon, Ga.: Mercer University Press, 2004.

Dormon, Caroline. "Blueprinting Wild Flowers." *Holland's Magazine,* June 1925.

———. "A New Voice from the Old South." *Holland's Magazine,* November 1926, 70.

———. "Southern Personalities: Lyle Saxon." *Holland's Magazine,* January 1931, 26+.

———. "Tree Studies in Blueprint." *American Forestry* (October 1924).

Dryer, Edward. Foreword to *The Friends of Joe Gilmore,* by Lyle Saxon. Gretna, La.: Pelican Publishing Co., 1998.

Dunn, Milton. "History of Natchitoches." *Louisiana Historical Quarterly* 3 (January 1920): 26–56.

Fairclough, Adam. *Race and Democracy: The Civil Rights Struggle in Louisiana, 1915–1972.* Athens: University of Georgia Press, 1995.

Faust, Ernest Carroll. "The History of Malaria in the United States." *Scientific American* 39, no. 1 (January 1951): 121–30.

Ford, Oliver. "François Mignon: The Man Who Would Be French." *Southern Studies* 2, no. 1 (Spring 1991): 51–59.

Keel, Bennie C., with Christina Miller and Mark Tiemann. *A Comprehensive Subsurface Investigation at Magnolia Plantation.* Tallahassee, Fla.: Southeast Archaeological Center, National Park Service, 1999.

Garber, D. D. "Mixed Architecture Adds Charm to Historic Melrose Plantation." *Dallas Morning News,* October 28, 1951.

Goodwin, William A. R. *Bruton Parish Church Restored and Its Historic Environment.* Petersburg, Va.: Franklin Press, 1907, 33. Accessed January 29, 2017. https://archive.org/stream/brutonparishchu01goodgoog#page/n8/mode/2up.

Goolsby, William Berlin. "Melrose Plantation Holds Open House for the Muse." *Dallas Morning News,* March 20, 1932.

Grainger, Ethel Holloman. "Miss Cammie Still Lives at Melrose." *Dallas Morning News,* March 20, 1932, 12.

Gutman, Lucy, and Shannon Frystak. "Carmelite 'Cammie' Garrett Henry: The Evolution of a Plantation Mistress and Chatelaine of the Arts." In *Louisiana Women: Their Lives and Times.* Edited by Mary Farmer-Kaiser and Shannon Frystak, 2:175–95. Athens: University of Georgia Press, 2016.

Harris, Bertha Cooper. *The Courage to Rise Again.* Austin, Tex.: Glad2b4u Press, 2012.

Harvey, Cathy Chance. "Dear Lyle / Sherwood Anderson." *Southern Studies* 18 (1979): 320–38.

Harvey, Chance. *The Life and Selected Letters of Lyle Saxon.* Gretna, La.: Pelican Publishing Co., 2003.

Haynie, Sandra Prud'homme. *Legends of Oakland Plantation: The Prud'hommes of Natchitoches Parish*. Shreveport: La. Press Co., 2001.

Hunter, Henley Alexander. *Magnolia Plantation: A Family Farm*. Natchitoches, La.: Northwestern State University Press, 2005.

Jackson, Joy. "Prohibition in New Orleans: The Unlikeliest Crusade." *Louisiana History* 19, no. 3 (1978): 261–84. Accessed September 25, 2016. http://www.jstor.org/stable /4231785.

Jackson, Sarah R. *Texas Woman of Letters: Karle Baker Wilson*. College Station: Texas A&M University Press, 2005.

Jacobs, Philip Walker. *The Life and Photography of Doris Ulmann*. Lexington: University Press of Kentucky, 2001.

Johnson, Fran Holman. *The Gift of the Wild Things: The Life of Caroline Dormon*. Lafayette: University of Southwestern Louisiana Press, 1990.

Johnson, T. R. "George Washington Cable." In KnowLouisiana.org, *Encyclopedia of Louisiana*. Edited by David Johnson. Louisiana Endowment for the Humanities, 2010–. Article published May 31, 2011. Accessed May 25, 2017. https://www.knowlouisiana .org/entry/george-washington-cable-2.

Jones, Anne Goodwyn. "*Gone with the Wind* and Others: Popular Fiction, 1920–1950." In *History of Southern Literature*. Edited by Louis D. Rubin Jr. et al., 363–74. Baton Rouge: Louisiana State University Press, 1985.

Kadlecek, Mabell R., and Marion C. Bullard. *Louisiana's Kisatchie Hills: History, Tradition Folklore*. Chelsea, Mich.: Sheridan Books, 1994.

Kane, Harnett T. *Plantation Parade*. New York: William Morrow, 1945.

Leary, Lewis. *William Faulkner of Yoknapatawpha County*. New York: Thomas Y. Cromwell, 1973.

Lee, Dayna Bowker. "Caroline Dormon: Louisiana's Cultural Conservator." In *Louisiana Women: Their Lives and Times*. Edited by Janet Allured and Judith F. Gentry. Athens: University of Georgia Press, 2009.

Lewis, Richard A. "Mr. Feuille." In *The New Encyclopedia of Southern Culture*, vol. 23: *Folk Art*. Edited by Cheryl Crown and Cheryl Rivers. Chapel Hill: University of North Carolina Press, 2013.

Louisiana Writers' Project. Works Progress Administration. *Louisiana: A Guide to the State*. Edited by Lyle Saxon. New York: Hastings House, 1941.

Lyons, Mary E. *Talking with Tebé: Clementine Hunter, Memory Artist*. Boston: Houghton Mifflin Harcourt, 1998.

MacDonald, Kevin C., et al. "The Archaeology of Local Myths and Heritage Tourism: The Case of Cane River's Melrose Plantation." In *A Future for Archaeology*. Edited by Robert Layton, Stephen Shennan, and Peter Stone, 125–42. Walnut Creek, Calif.: Left Coast Press, 2006.

MacNeil, Joe. "It Doesn't Matter Where You Work." *New York Observer*, June 3, 2016. Accessed June 15, 2016, http://observer.com/2016/06/it-doesnt-matter-where-you-work/.

Maxwell, Allen. "*Children of Strangers* by Lyle Saxon." *Southwest Review* 23, no. 1 (October 1937). 117–20. JSTOR (43166420).

Meese, Elizabeth. "What the Old Ones Know: Ada Jack Carver's Cane River Stories." In *Louisiana Women Writers: New Essays and a Comprehensive Bibliography*. Edited by Dorothy H. Brown and Barbara C. Ewell. Baton Rouge: Louisiana State University Press, 1992.

Megraw, Richard B. *Confronting Modernity: Art and Society in Louisiana*. Jackson: University Press of Mississippi, 2008.

"Melrose Auction Sale, June 6 and 7, 1970. A Report by Northwestern State University of Louisiana." Association for the Preservation of Historic Natchitoches, Natchitoches, La., April 2011. Accessed March 10, 2014. http://www.aphnnatchitoches.net/page/428625602.

"Melrose Plantation Long-Range Interpretive Plan." Association for the Preservation of Historic Natchitoches, Natchitoches, La., 2013.

Mencken, H. L. "The Sahara of the Bozart." In *The American Scene: A Reader*, by H. L. Mencken. Edited by Huntington Cairns, 157–68. New York: Alfred A. Knopf, 1977. Accessed January 15, 2016.

Mignon, François. *Plantation Memo: Plantation Life in Louisiana, 1750–1970, and Other Matter*. Edited by Ora Garland Williams. Baton Rouge: Claitor's, 1972.

Mills, Elizabeth Shown. "Demythicizing History: Marie Thérèse Coincoin, Tourism, and the National Historical Landmarks Program." *Louisiana Historical Quarterly* 53, no. 4 (Fall 2012): 402–37.

Mills, Gary B. *The Forgotten People: Cane River's Creoles of Color*. Rev. ed. by Elizabeth Shown Mills. Baton Rouge: Louisiana State University Press, 2013.

Moore, Diane M. "Mistress of Melrose: Carmelite Henry, 1883–1948." In *Their Adventurous Will: Profiles of Memorable Louisiana Women*, 3–31. Lafayette, La.: Acadiana Press, 1984.

Moore, Edith Wyatt. *Natchez under the Hill*. Natchez, Miss.: Southern Historical Publications, 1958.

Morel, Vera. "Old Louisiana Memories Are Kept Green in Miss Cammie's Library." *New York Sun*, August 11, 1937.

Morgan, David W., Kevin C. MacDonald, and Fiona J. L. Handley. "Economics and Authenticity: A Collision of Interpretations in Cane River National Heritage Area, Louisiana." *George Wright Forum* 23 (2006): 44–61.

Nardini, Louis Raphael, Sr. *My Historic Natchitoches, Louisiana, and Its Environment*. Natchitoches, La.: Natchitoches Publishing Co., 1963.

Nichols, Irby C. "Announcement of Natchitoches Meetings March 13, 14." *Mathematics News Letter* 5, no. 5 (January 1931): 5–7. JSTOR (3027931).

Osterweis, Rollin G. *The Myth of the Lost Cause, 1865–1900*. Hamden, Conn.: Archon Books, 1973.

Parrish, Susan Scott. *The Flood Year, 1927: A Cultural History*. Princeton: Princeton University Press, 2017.

Powell, Lawrence N. *New Masters: Northern Planters during the Civil War and Reconstruction*. New Haven: Yale University Press, 1980.

Rawson, Donald M. "Caroline Dormon: A Renaissance Spirit of Twentieth-Century Louisiana." *Louisiana History* 24, no. 2 (Spring 1983): 121–39. Accessed March 8, 2014. http://www.jstor.org/stable/4232260.

Reed, John Shelton. *Dixie Bohemia: A French Quarter Circle in the 1920s*. Baton Rouge: Louisiana State University Press, 2012.

Rubin, Richard. "The Colfax Riot." *Atlantic*, July–August 2003. Accessed October 27, 2017. https://www.theatlantic.com/magazine/archive/2003/07/the-colfax-riot/378556/.

Saxon, Lyle. "Cane River." *Dial* 80 (1926): 207–21.

———. "The Centaur Plays Croquet." In *The American Caravan*. Edited by Van Wyck Brooks, Alfred Kreymborg, Lewis Mumford, and Paul Rosenfield, 344–69. New York: Literary Guild, 1927.

———. *Children of Strangers*. Boston: Houghton Mifflin, 1937.

———. "Easter Sunday at Aunt Cammie's." *New Orleans Times-Picayune Magazine*, May 20, 1923, 4, 7.

———. *Fabulous New Orleans*. Gretna, La.: Pelican Publishing Co., 2004.

———. *Father Mississippi: The Story of the Great Flood of 1927*. Gretna, La.: Pelican Publishing Co., 2006.

———. *The Friends of Joe Gilmore*. Gretna, La.: Pelican Publishing Co., 1998.

———. *Lafitte the Pirate*. New York: Century Co., 1930.

———. "The Long Furrow." *Century Magazine* 114 (1927): 688–96.

———. "The Mistletoe Trail: Christmas on a Louisiana Plantation." *New Orleans Times-Picayune Magazine*, December 23, 1923.

———. *Old Louisiana*. Gretna, La.: Pelican Publishing Co., 1998.

Scarborough, Dorothy. *On the Trail of Negro Folk Songs*. Cambridge: Harvard University Press, 1925.

Schuyler, Lorraine Gates. *The Weight of Their Votes: Southern Women and Political Leverage in the 1920s*. Chapel Hill: University of North Carolina Press, 2006.

Scott, Anne Firor. *The Southern Lady: From Pedestal to Politics, 1830–1930*. Chicago: University of Chicago Press, 1970.

Scott, John W. "William Spratling and the New Orleans Renaissance." *Louisiana History*

15, no. 3 (Summer 2004): 387–399. Accessed September 1, 2017. http://www.jstor.org
/stable/4234032.

Seebold, Herman. *Old Louisiana Plantation Homes and Family Trees*. Baton Rouge. Pelican
Publishing Co., 2005.

Shiver, Art, and Tom Whitehead. *Clementine Hunter: Her Life and Art*. Baton Rouge: Louisiana State University Press, 2012.

Smith, Armantine M. "The History of the Women's Suffrage Movement in Louisiana."
Louisiana Law Review 62, no. 2 (Winter 2002): 509–60. http://digitalcommons.law
.lsu.edu/cgi/viewcontent.cgi?article=5926&context=lalrev.

Smith, J. Frazer. *White Pillars: The Architecture of the South*. New York: Bramhall House,
1941.

Snell, David. "The Green World of Carrie Dormon." *Smithsonian*, February 1972, 28.

Spratling, William, illus., and arranged by William Faulkner. *Sherwood Anderson and
Other Famous Creoles: A Gallery of Contemporary New Orleans*. New Orleans: Pelican
Bookshop Press, 1926.

Stanonis, Anthony J. *Creating the Big Easy: New Orleans and the Emergence of Modern Tourism, 1918–1945*. Athens: University of Georgia Press, 2006.

Stuart, Bonnye E. *More than Petticoats: Remarkable Louisiana Women*. Guilford, Conn.:
Morris Book Publishing, 2009.

Thomas, James W. *Lyle Saxon: A Critical Biography*. Birmingham, Ala.: Summa Publications, 1991.

———. "Lyle Saxon's Struggle with *Children of Strangers*." *Southern Studies* 16, no. 1
(Spring 1977): 27–40.

Toth, Emily. *Unveiling Kate Chopin*. Jackson: University Press of Mississippi, 1999.

Tyler, Pamela. "Woman Suffrage." In KnowLouisiana.org, *Encyclopedia of Louisiana*. Edited by David Johnson. Louisiana Endowment for the Humanities, 2010–. Article
published March 8, 2016. Accessed September 24, 2017. http://www.knowlouisiana
.org/entry/woman-suffrage.

Wells, Carol. "Agnes Morris." *Louisiana History* 27, no. 3 (Summer 1986): 261–72.

———. *Cane River Country*. Natchitoches, La.: Northwestern State University Press, 1979.

White, Alice Premble. "The Plantation Experience of Joseph and Lavinia Erwin, 1807–
1944." *Louisiana Historical Quarterly* 27, no. 2 (April 1944): 343–477.

Williams, Ora G. "Cammie G. Henry." In *Four Women of Cane River: Their Contributions to
the Cultural Life of the Area*, 19–25. Natchitoches, La: Natchitoches Parish Library, n.d.

Williams, Susan Millar. *A Devil and a Good Woman, Too: The Lives of Julia Peterkin*. Athens:
University of Georgia Press, 1997.

Wilson, Charles R., and William Ferris. *Encyclopedia of Southern Culture*. Chapel Hill:
University of North Carolina Press, 1989.

Wilson, Edmund. *The Twenties.* New York: Farrar, Straus, and Giroux, 1975.

Wilson, Mary Ann. "Grace King." In KnowLouisiana.org, *Encyclopedia of Louisiana.* Edited by David Johnson. Louisiana Endowment for the Humanities, 2010–. Article published June 3, 2011. Accessed May 25, 2017. http://www.knowlouisiana.org/entry/grace-king-3.

DISSERTATIONS AND THESES

Ford, Oliver Jackson. "Ada Jack Carver: A Critical Biography." Ph.D. diss., University of Connecticut, 1975.

Luster, Sarah Bailey. "The Natchitoches Art Colony: A Southern En Plein Air Art Colony." Master's thesis, Tulane University, 1992.

Olson, Jane Sharrett. "Stephen Garrett Henry: The Soldier from Melrose." Master's thesis, Northwestern State University of Louisiana, May 1, 1990.

Smith, Jenny. "An Examination of the Relationships between a Literary Patron and Her Son and Daughter." Senior thesis, Louisiana Scholars' College at Northwestern State University of Louisiana, 1992.

Taylor, Donnis Marie Ward. "Louisiana's Literary Legacy: A Critical Appraisal of the Writings of Ada Jack Carver." Ph.D. diss., University of Southwestern Louisiana, 1991.

Thorn, Laura. "A Place of Stillness: Melrose and the Southern Literary Community of the 1920's and 1930's." B.A. thesis, Louisiana Scholars' College, Northwestern State University of Louisiana, 1990–91.

INDEX